Springer Series in Advanced Manufacturing

Series Editor

Duc Truong Pham

For further volumes:
http://www.springer.com/series/7113

Marco Gobetto

Operations Management in Automotive Industries

From Industrial Strategies to Production Resources Management, Through the Industrialization Process and Supply Chain to Pursue Value Creation

Springer

Marco Gobetto
Automotive Engineering Faculty
FIAT Sepin S.cp.A. and Politecnico of Turin
Turin
Italy

ISSN 1860-5168 ISSN 2196-1735 (electronic)
ISBN 978-94-007-7592-3 ISBN 978-94-007-7593-0 (eBook)
DOI 10.1007/978-94-007-7593-0
Springer Dordrecht Heidelberg New York London

Library of Congress Control Number: 2013947779

Printed on acid-free paper

Springer is part of Springer Science+Business Media (www.springer.com)

If you think of standardization as the best that you know today, but which is to be improved tomorrow; you get somewhere

Henry Ford

Foreword

With more than 30 years of experience in the automotive industry, eight of those involved in the development and implementation of World Class Manufacturing, and having, over the years, experienced almost all the operation sectors through all economic scenarios, I can state, without a doubt, that a wise Management of Operations is the key to company success.

For this reason, the technical background of a person who works in, or aims to join, the automotive sector, should include Operations Management (OM). We therefore offer this book as an answer to the need for a repository of detailed knowledge of this topic.

Writing a book on OM is as ambitious as it is critical, since it is required to deal with all processes and aspects of the automobile lifecycle management Marco Gobetto was the perfect candidate for this task. Strengthened by working experience in different operation sectors of the Fiat Group and a teaching background at Politecnico of Turin, he succeeded in embracing all aspects of OM in great detail.

It is for these reasons that I encouraged and sustained him in this enterprise and am proud to write this Foreword.

Today, companies are competing in a much more complex environment than it was the case only a few years ago. Survival rather than growth is the immediate goal. They must focus on quality, time-based competition, efficiency, international perspectives, and customer satisfaction. Competition driven by a global market, e-business, the Internet, and advances in technology all require flexibility and responsiveness. This new focus has placed OM in the limelight of business, because it is the function through which companies can achieve a necessary level of competitiveness.

The automotive industry is and will remain to be one of the key industries of the world. If auto manufacturing were a country, it would be the sixth largest economy, with over 66 million cars, vans, trucks, and buses manufactured (data from OICA, in 2005). These vehicles are essential to the working of the global economy and to the wellbeing of the world's citizens. This level of output is equivalent to a global turnover (gross revenue) of almost €2 trillion.

Traditionally, the automotive industry was organized by rigid labor division and paced assembly lines were employed to realize low-cost mass production. Nowadays, flexible mixed model assembly lines make it possible to produce a large variety of customized products. As an example, Mass Customization is one of the answers to flexibility challenge: the ability of a firm to highly customize its goods and services to different customers. Mass customization requires designing flexible operations and using delayed product differentiation, also called postponement. This means keeping the product in generic form as long as possible and postponing radical change.

This is not for free. Increasing flexibility in design and manufacturing imposes new challenges not only for development of the product and manufacturing system, but also for the logistics coordination within global supply chains, purchasing, and marketing operations. In addition, novel planning and scheduling approaches are needed in order to manage the mismatch between increased product variety and the need to improve utilization of capital-intensive resources.

In other words an Operations Management, to be successful, has to focus on adding value during the transformation process. By value added in this context we mean the net increase between the final value of a product and the value of all the inputs. The greater the value added, the more productive a business is.

An obvious way to add value is to reduce the cost of activities in the transformation process by finding and eliminating waste and losses. In addition to value added, operations must be efficient, that is, being able to perform activities well, and at the lowest possible cost. All activities must be analyzed to eliminate those that do not add value, and restructure processes and jobs to achieve greater efficiency.

In this way, Operations Management has become the focal point of efforts to increase competitiveness by improving value added and efficiency for the whole company.

Achieving these targets requires both strategic and tactical management. Strategic management involves making decisions about what business the organization will be in and what its overall objectives will be. Strategic management's planning is long term and considers where the business wants to be in 2–3 years, or longer in some cases. Tactical management involves making decisions about how an organization should go about achieving the overall objectives determined by strategic management. Tactical management decides what needs to be done within that year to implement the plan of strategic management. Tactical decisions must be aligned with strategic decisions, because they are the key to the company's effectiveness in the long run. Tactical decisions provide feedback to strategic decisions, which can be modified accordingly.

Having a well-coordinated management system focused on adding value and improving efficiency could be enough. Today, businesses must think in terms of competition based on timing and a global marketplace.

This includes developing new products and services faster than the competition, reaching the market first, fulfilling customer orders most quickly, meeting customer needs, and getting the right product to markets as diverse as the Far East, Europe, and Africa.

Operations Management is responsible for most of these decisions. OM decides whether to tailor products to different customer needs, where to locate facilities, how to manage suppliers, and how to meet local government standards.

This book copes with all aspects of Operation Management, reflecting the ideal of Mutually Exclusive, Collectively Exhaustive (MECE), which I consider as the basic requirement to be World Class.

In Chap. 1, after an enjoyable overview on automotive history, the reader will discover information on all processes and technologies involved in an automotive enterprise and learn why decision making is a key factor of success. The chapter highlights method, tools, criteria, and knowledge which support a decision-making process guaranteeing a high level of success.

Chapters 2 and 3 deal, in a wealth of details, with the development process of product and manufacturing technologies and methods. Flexibility premises are generated in this phase. The theme of producing a large variety of customized products by maintaining a high level of product and process standardization is covered by explaining all the meaningful methodologies, requirements, and parameters managed by Engineering Departments. A special highlight emphasizes Information Technology which, nowadays, allows effective application of Concurrent Engineering to coordinate different departments in different parts of the world.

From Chaps. 4–8, the reader will find a detailed analysis of the new methodologies to improve productivity and to sustain continuous improvement. More than other chapters, this part has my personal appreciation. Marco has demonstrated his capability in having successfully dealt with all the aspects that an Operations Manager should know and easily manage: the four pillars of a Production System, Workplace Organization, Logistics, Professional Maintenance, and Quality. I personally urge you to ponder carefully all the aspects contained in these four chapters, since they contain the essential knowledge of a company needed to survive in the future.

The last chapter is about value and how it is generated. As said above, Operations Management plays the most important role in increasing competitiveness by improving value and efficiency for the whole company, since all activities or initiatives should be in the direction of creating value.

For all people who want to participate in or deeply understand the automotive industry, I would foster reading this book to deeply catch the complexity and the continuous challenges of the automotive enterprise. To all Automotive Engineering students, my wish is that you will be guided by this book and will keep it in your personal library as a knowledge reference for the future.

Luciano Massone

Acknowledgments

The list of people I would like to thank is quite long because of the extended period required to complete, and finally release, this book with coverage of practically a decade of blended working experiences, a mix of teaching at the University and professional worldwide working in Manufacturing through many different Industrial Businesses in a continuously changing scenario.

So I would like first of all to thank Carlo Mangiarino, experienced Manager through most of the companies in FIAT; without him it would not have been possible for me to have the chance to start this exciting experience at the University; experience in which I strongly believed from the beginning, while combined with my actual professional assignment in FIAT. From Carlo I learned not only the professional side but also a unique human style.

A special thanks goes to Domenico Maisano and Luca Mastrogiacomo, both brilliant young engineers and doctors in research at DISPEA Department in Politecnico of Turin; Domenico started the collaboration with Production Management from the beginning and specifically dedicated himself to writing the Appendix for the basics of production management; Luca continued brilliantly to assist me during the growth of the international course and has been invaluable in refining this part of the text and improving it during these last 3 years; to them my best wishes for brilliant international careers at the University of Turin.

I would like to thank Giancarlo Genta, Professor at Politecnico of Turin and Responsible of Automotive Engineering Course, for his commitment to the internationalization of the courses in this Faculty.

A great thank you goes to Luciano Massone, Vice President of World Class Manufacturing in FIAT Group Automobiles and a Member of the Steering Committee for Automotive Engineering, a committee shared between FIAT and Politecnico of Turin. I met Luciano for the first time in 2005 and he was a "stand alone" man at that time in his new department; he asked me and another few colleagues of mine to help him with a special initiative that aimed to build the New FIAT Production System. This initiative, only a few years later, contributed significantly to a great turnaround in FIAT at worldwide level and to the Joint Venture between FIAT and Chrysler. World Class Manufacturing is currently more than simply a new production system for FIAT sectors and plants, but almost a common language to share best practices and give new opportunities to the companies that want to join this way of business life. Now Luciano leads a strategic group of people that possesses a body of knowledge considered as one of the main assets of our company: thank you for offering me this “fast pass” to learning, experiencing, and growing up professionally in such a very short time.

Another special thank you goes to Elena Sacerdote, before FIAT Corporate Learning Officer and now FIAT Brand Customer Care Manager. From the beginning of her assignment she gave the right relevance to her Manufacturing Training Role and, with her commitment, contributed significantly in keeping it close to the Manufacturing Central Department in FIAT, ensuring always the right link with central training logics as well. Dealing with training daily, I have learned how difficult it is to make other people understand how strategic it is; Elena’s work lives in the pages of this book.

Last, but not the least, a devoted “thank you” to all my Family, for supporting me in all these intense working years and giving me the opportunity to nourish this knowledge: my wife Arianna for being patient and for her continuous support year after year, my daughter Gaia for the motivation she is able to induce in me, my father and my mother for having given me the chance to be here at this point.

A final thank you goes to all those readers who will use this knowledge as a starting point, will apply it to their professional life, and will increase their curiosity to learn more on some of the chapters treated in this book.

Marco Gobetto

Contents

Abbreviations

A	Technical Availability
AFC	Annual Fixed Cost
AM	Autonomous Maintenance
APC	Available Productive Capacity
APP	Aggregated Production Planning
AT	Active Time
ats	Average technical saturation
AUT	Available Up Time
AWT	Available Working Time
BM	Breakdown Maintenance
BOM	Bill Of Material
BUT	Breakdown Unavailable Time
CAD	Computer Aided Design
CAE	Computer Aided Engineering
CAPE	Computer Aided Process Engineering
CBM	Condition-Based Maintenance
COS	Cost Of Sale
DAV	Developed Activity Volume
DLR	Direct Labour Requirement
DR	Design Review
ECUT	External Causes Unavailable Time
EOQ	Economic Order Quantity
FI	Flow Index
FIFO	First In First Out
FMEA	Failure Mood Effect Analysis
FMECA	Failure Mood Effect and Causes Analysis
FOQ	Fixed Order Quantity
HDM	Hours of Direct Manpower
HIM	Hours of Indirect Manpower
HVP	Hourly Virtual Production or Productivity
IAA	Individual Activity Achievable
ILP	Individual (annual) Labour Productivity
IT	Idle Time
IWH	Individual Working Hours

JIS	Just In Sequence
JIT	Just In Time
KAI	Key Activity Indicator
KMI	Key Management Indicator
KPIs	Key Performance Indicators
LCA	Low Cost Automation
LCV	Light Commercial Vehicle
LIFO	Last In First Out
LT	Lead Time
LTP	Lot Turnover Period
MMUT	Manpower Management Unavailable Time
MP	Monthly Production
MPS	Master Production Scheduling
MPV	Multi-Purpose Vehicle
MRP	Material Requirement Planning
MTBF	Mean Time Between Failures
MTTR	Mean Time To Repair (and Restart)
NAFTA	North America Free Trade Area
NVAA	Non Value Added Activities
OEE	Overall Equipment Efficiency (or Effectiveness)
OELT	Order Execution Lead Time
OP	Operative Plan
OPF	One Piece Flow
PAV	Planned Activity Volume
PBS	Product Breakdown Structure
PDCA	Plan Do Check Act
PDM	Product Data Management
PFD	Product Function Deployment
PLT	Process Lead Time
PMS	Product Macro Structure
PMUT	Preventive Maintenance Unavailable Time
pu	Process unavailability index
PWT	Planned Working Time
QR	Quality Requirements
R	Technical Reliability
SC	Supply Chain
SCM	Supply Chain Management
SFC	Shift Fixed Cost
SL	Service Level
SPC	Standard Productive Capacity
ST	Standard Time
SOT	Standard Operation Time
SUUT	Set-Up Unavailability Time
SUV	Sport Utility Vehicle
TBM	Time-Based Maintenance

TPM	Total Productive Maintenance
TQM	Total Quality Management
TT	Takt Time
TTC	Total Transformation Cost
TTM	Time To Market
TUC	Transformation Unitary Cost
U	Utilization (degree)
UST	Unitary Set-Up Time
UVC	Unitary Variable Cost
VAA	Value Added Activities
VTC	Variable Transformation Cost
WBS	Work Breakdown Structure
WCT	Working Cycle Time
WIP	Work In Progress
WTE	Working Time Employed

Introduction

This book is the outcome of a very long and intensive project begun in 2002, when Carlo Mangiarino, longstanding Manager for all the major companies of FIAT and then head of the course of Production Management for the Faculty of Automotive Engineering at the Politecnico of Turin, asked me to collaborate with him in the development of this course, both for the content and for the lectures at the university.

We began designing this textbook in Italian, since that was the language that the course was in, and then, year after year, updated and integrated it with fresh material according to the evolution of the techniques and scenarios in the automotive industry.

In 2007, when I took over stewardship of Production Management from Carlo, the course expanded into other languages. Based additionally on my experience worldwide as Manager of a training unit for manufacturing disciplines for FIAT, I began developing the English language version, revisiting the entire book in order to integrate all the information to come out of the automotive industry in recent years.

So far, this book serves mainly as a text reference for the course of Production Processes, Safety, Organization, and Management I run at the Faculty of Automotive Engineering, but at the same time it represents a good compendium of basics on Industrial Management, since it covers all major chapters: from product development and "make or buy" decision strategies to the setting of manufacturing systems and management, through analysis of the main resources needed for production, and finally, exploring management of the supply chain and procurement techniques. The very last chapter is dedicated to summing up all topics through analysis of key management indicators for pursuing the value creation at the heart of every industrial business.

A portion of the Appendix is dedicated to the basics of production management where all main relevant definitions, techniques, and criteria are addressed through the inclusion of numerical examples, for the purpose of providing the knowledge needed to approach all other chapters.

It is my hope that this book will lead both Automotive Engineering students and also a wider range of readers to a better understanding of the automotive world and its industrial aspects.

Marco Gobetto

Chapter 1
Historical Outlines and Industrial Strategies for Automotive Industries

1.1 Historical Outlines in the Automotive Industry

We will briefly examine the evolution of industrial strategies for the development of cars and commercial vehicles during the entire historical period from the birth of the car to the present day. We will deal particularly with manufacturing engineering, work analysis and international development plans adopted by different companies throughout this time.

The key changes in industrial strategies are, from one side, the evolution of market and vehicle usage conditions and, from the other, the availability of new technologies, and safety and environmental restrictions. Competition between competitors leads to different speeds of development at different times.

Considering social, economic, technological and managerial developments, we can break down the assorted historical periods as follows:

A. ***1885/1905—Pioneers and the birth of big brands***
   Historically, the birth of the car began in Europe and it would characterize the entire 19th century from the point of view of innovation. The first cars were "motorized coaches" and they were produced with craft methods, the operative instructions for assembling the car transforming engineers directly into craftsmen. These vehicles were plagued by many functional defects that could only be fixed after long sessions of trial and error on the road conducted by expert mechanics.
   In this "pioneer" phase. several techniques and schools of craft were affirmed in Europe; the best engine solutions came from Austria and Germany, while space frame and body solutions came from France and Italy; the English style would be distinguished for its luxury and would even end up influencing public transportation solutions.
   By the introduction of the first automatic machine tools, the method of production had improved; furthermore, new tools for geometrical measurement allowed for better measuring of mechanical and body in white elements. It is in this period that the first research centre began to operate in collaboration with

M. Gobetto, *Operations Management in Automotive Industries*,
Springer Series in Advanced Manufacturing, DOI: 10.1007/978-94-007-7593-0_1,

Technical and Scientific Universities with the aim of producing with a high level of consistency: this was the transformation of automotive production from an artisanal method to an industrial one, leading to the concept of standard and serial production.

It was in the first decade of the last century that the first touring cars were built; for these vehicles, a stylish shape and high quality materials became very important.

In Europe, during the so-called "bell' époque" period, a lot of small and medium carmakers began work; many of their products were selected by very demanding customers (almost exclusively the nobility and middle class) who continued to be interested in technical innovation and original solutions. Some of these cars also proved successful in competitions.

Conversely, in the United States of America, likely due to the necessity of traveling long distances on rough territory, cars were considered to be more for practical and functional purposes; there were fewer carmakers than in Europe, and yet more opportunity for finding capital resources for investment.

B. ***1905/1925—Mass production: Ford strategic model***

Henry Ford understood that the car could become a mass product with the assumption of certain fundamental features such as reliable functionality and high versatility. The real key to success lay in creating a rapid production capacity to satisfy increasing demand and thus win the market competition.

To do that, Ford combined his idea with F.W. Taylor's model for the *scientific organization of work*:

- re-playable and interchangeable elements of the product;
- production organized by method and predefined working times;
- simplification of job tasks by specialization;
- blue collar motivation to speed-up production in a short period of time.

The real advantage of this model was the ease with which it could be introduced into the production process, lowering the cost of manpower in the workforce and proving easy to establish anywhere. While processes such as steel printing and metal forging required capital intensive solutions such as the introduction of big machinery, in the final assembling area, the product could be conveyed by a continuous flow line, progressively adding materials and components at different stations at which workers followed standard operating procedures. It was the beginning of the *assembly line*.

Metal cutting and machining were organized in a special unit, by using high cadence equipment and production organized in batches. As a consequence of these solutions, big warehousing areas became necessary, the management of which demanded the establishment of a *material management* department. The logistic management was aided by data processing systems available at that time.

To overcome the logistic difficulties and the change of management with the suppliers, Ford primarily concentrated the production process at its own Detroit-Rive Rouge plant, reaching the production rate of 1 million vehicles a year. This same prerogative would come to be the weak point of the Ford strategy in the future.

The success of the Model "T", produced in a number of different styles on the same frame and with the same engine, allowed Ford to reach 60 % of the American market and maintain 50 % of it up to 1920. These results would be the impulse for other carmakers, including those in Europe, to adopt similar strategies for production systems, even if volume of production wasn't comparable.

The First World War stopped the car industry's development and gave a big boost to the industrial vehicle industry (agricultural and transport), while the production capacity used for military vehicles would soon be converted for the manufacture of civic vehicles, though on a lower production scale.

It is undeniable that, at the end of this period, certain special technologies (tires, special steels…) developed for military application and low fuel costs contributed to the evolution of the car's functionality (Fig. 1.1).

C. ***1925/1965—The 2nd change in management: product diversification orienting and plant specialization***

At the beginning of the 1920s, Ford's strategies began to show some limitations as soon as customers started searching for more creative offerings and more customized products, including features not found on Ford cars.

In 1921, General Motors, owner of the Chevrolet, Cadillac, Oldsmobile, Buick

Fig. 1.1 A "Model T" at the ford museum in Dearborn, Michigan

and Pontiac brands, under the leadership of its third president A.P. Sloan, developed a new strategy: building series cars with a wider product range depending on a customer's economic needs.

The product diversification theme was developed by GM without utilizing F.W. Taylor's organizational criteria from the previous decade. Plants were established by brand and car class and, at the same time, scale synergies were set for "capital intensive" productions: foundry and forging, steel printing, powertrain systems manufacturing, electrical systems, etc. All of these activities were controlled by the Holding Company while production was allocated to specific plants connected by railways to the final assembling plants.

Practically, A.P. Sloan began relocating manufacturing and managerial processes, assuring central financial and economic control. Plants were now not only cost centres but also profit centres that were called to provide their own investments.

With this new industrial strategy, General Motors surpassed Ford in market shares, increasing cash flow for use in the increase of production capactity and expansion of the dealer network, as well as improved after-market technical assistance. Only some years later, after the "big depression" in the 1920s, did Ford recover its position in the car market by rethinking its strategy, ending production activities in Rive Rouge, a plant that was too high in equipment density and too big to manage.

In Europe, especially Italy, Germany and France, mass production of a super-economical car was the leading concept, even if it did not come into complete fruition until after the Second World War.

In fact, during the 1950s, in the middle of the economic rebirth, European carmakers began following Ford's strategic model in order to ramp-up mass production as quickly as possible, so much so that certain specific equipment was imported directly from Michigan and Illinois or built in Europe on American license.

European cars are somewhat different from American cars: the mass-produced cars are small sized and more agile, while the top class cars are faster and well refined. The frame is linked to the front wheels and the body is self-supporting (the Lancia concept). They employ a front traction solution, proposed first by Citroen and later on refined by FIAT in mass production with the powertrain system cross-assembled. The engines exhibit higher performance thanks to very sophisticated carburetion systems with a valve. (These solutions would later be adopted by major European and Japanese carmakers once their manufacturing solutions were adapted to big series production.)

D. ***1965/1980—International diffusion of mass production; oil crisis and customer protection policies***

Ford and General Motors exported their industrial management model worldwide, increasing their business considerably. Furthermore, they developed important production capacities both in Western Europe and then South America and the Far East.

Even European carmakers followed the strategies of international market development: they tried to export a large quantity of cars to North America and built new plants in emerging countries by CKD (Completely Knocked Down) systems. Fiat and Renault were particularly involved in URSS and COMECON countries in supplying specific know-how and equipment for mass series production.

In the meantime, Japanese carmakers had begun using solutions for production systems similar to the ones proving successful in the Western countries, increasing their production capacities year by year, beginning with internal and South-East Pacific markets. Throughout the 1970s, they were able to improve their standard such that exportation to Europe and North America correspondingly improved.

To achieve these targets, they instituted a system of strenuous labor far beyond anything found in North America and Europe, both of which had workers who believed they were entitled to a life outside of work.

In the same period, some important centres of engineering and style model design were established in Italy, many dedicated to special designs for sports and luxury cars (Pininfarina, Bertone, Ghia… and later Italdesign). Style design and functional features, such as vehicle aerodynamics, were developed with scientific methodology and experimental techniques that allowed for significant results on both the technical and style sides.

These solutions would later be used for mass series production by even the biggest carmakers worldwide.

By the early 1970s, after a decade of growth, the bigger carmakers reached the maximum level of employment in their historical plants and their influence on socio-economic development in industrial countries was very high. In Europe at the end of the 1960s, the unions and general social upheaval limited the increase of production volumes (especially in England, France and Italy). To overcome the problem of "job disaffection", the Tayloristic model needed to be revisited, while in Germany and Japan the industrial relationship model remained constant in accordance with productivity and industrial growth.

During 1973/1974, after a long period of continuous improvement, the automotive industry, particularly in Europe, felt the shock of the big oil crisis. Furthermore, the "*customer protection*" and "*emission and energy saving*" policies approved in the USA became a big constraint on the ability to revisit product development criteria and solutions, especially for powertrain and exhaust systems, so that different solutions were needed for North America's car market than those being developed in Europe.

After the second oil crisis in 1979, some of the major names in the automotive industry (Daimler-Benz, Fiat, Saab-Scania) began to differentiate their industrial activities to minimize the effects of the car market on economic results.

Regarding the evolution of "manufacturing systems", up to the mid-70's, equipment was mostly based on *hard automation* and work on fixed cycles. For each model changeover, big investment and long periods of production stoppage were needed, leading to a delay in the launch of new products. The life

cycle of the product was up to 12/15 years for American carmakers: thus, they used the "model year" policy only to indicate certain external redesigns.
In the subsequent years, the introduction of "programmable production systems" became strategic (Numeric Control Machines, Multi-operational Machine Centres, Robot Equipment) together with computer systems based on digital technologies—developed in the USA—with the ability to improve the management of logistic systems.
During the same period, the European machine tool industry not only improved, but did so enough to overcome its American counterpart, as it did again a few years later to the Japanese. The first generation of flexible production systems, so-called CIM (computer integrated manufacturing system), was born in this period, but, due to a complex level of technology, these systems proved difficult to manage.
In Japan, Toyota followed a completely new industrial strategy, the principal aim being product quality, lean production processes and personal involvement. Kiichiro Toyoda (Toyota founder) gave to Taiichi Ohno the task of developing effective methodologies for preventing design and manufacturing errors. Professional skills were developed through a rigorous "company school". All operative employees were involved in continuous improvement (*kaizen*) through teamwork.
This organizational model was successful and quickly spread throughout the Japanese industry. Before long Japanese carmakers had become highly competitive, obtaining more and more shares of the American car market by the day and challenging the three "big companies" (GM, FORD and CHRYSLER) during the final part of this historical period.

E. ***1980/1995—New strategies for cooperation between Final and Middle Producers. Japanese organizational model diffusion: "total quality" and "lean production"***
With the '80s began a new age of progress in the automotive industry. The final product became more complex from a technological point of view, thanks, in part, to ecological and safety restrictions and also to customer requests regarding functionality and comfort. The ability to integrate the electronic systems for engine control, safety and air conditioning represented a key factor for success.
For the transportation of goods, a new generation of high capacity vehicles with more severe safety restrictions and availability was developed.
For this purpose, component industries assumed a progressively more important role than in the past, transforming from the producers of simple "on design" parts into the developers of functional product sub-systems, based on the carmakers' guidelines.
The model of cooperation between the final and middle producers was structured in different strategic profiles, based on a macro-industry of necessity:

(a) **The German model** promoted local suppliers into leaders of R&D activities, reserving priority application of the innovations for German carmakers (for example, the strategic alliance of innovation between Bosch and Mercedes);
(b) **The Japanese model** led carmakers to support the supply chain as the first level of R&D capacity, linking them through company capital investment participation (for example, Toyota controlling Denso) rather than promoting funds without any rate of interest;
(c) **The American model,** commercially more open and oriented toward obtaining maximum economical results in a short period, led carmakers to select a supply chain concentrating on independent suppliers with market leadership; these were then engaged in developing a worldwide production system, decentralizing labour intensive processes into countries with low manufacturing costs.

For other industrial compartments, relationships between carmakers and their suppliers tended mostly towards the American model.

Due to a greater complexity of product, most carmakers used specific suppliers for the manufacturing of specific parts and the engineering and installation of specific equipment. This strategy is called "outsourcing" and helped plants to reduce complexity of management in internal processes. Production of high technological components was concentrated within a few leading market suppliers.

During the 1980s, the major carmakers, in particular, Volkswagen, Fiat and Volvo, would use hard automation, with major investments into the increase of productivity and improvement of the working environment. Flexible and multi-purpose manufacturing systems were used in moderation to effect a quick changeover of the product range with less specific investments. Even ICT systems were used.

Numerically controlled machine tools, equipped with auto compensation systems, helped in achieving very high precision: 5000 mm for the joining of kinematic parts and 1/500 of an mm for the joining of structural vehicle parts. Furthermore, high resistance steel zinc-coated and anticorrosion coating let engines and vehicles greatly extend their technical life.

New, more interesting and safer car models generated new needs in the car market. So a new market was born: the second-hand car market. Government facilitation in Western Europe, oriented towards reducing environmental pollution, helped customers in buying a new car. As a consequence, the second-hand used market became more important.

At the beginning of the 1990s, the most relevant industrial event would be the so-called "transplant strategy", in which final assembly plants were established by Japanese carmakers in North America and Western Europe, increasing market shares in these markets.

During this historical period, even in western companies, the Japanese manufacturing model began to spread. The main features of the so-called *Total Quality Management* and *Lean Production* are:

1. Total quality orientation, obtained by focusing on the final customer and by applying *continuous improvement activities;*
2. Involvement of people in the product development department by using a *teamwork approach;*
3. Reduction of "time to market" by leading in parallel product design, prototyping and manufacturing engineering activities (*simultaneous engineering*);
4. Simplification of logistic processes on the shop floor (*just in time, kanban…*);
5. Application of modular manufacturing systems (*agile systems*);
6. Maximum equipment utilization and *minimum unit production cost*, assuming smoothness of production.

   This model, originating with Toyota, was approached progressively by several carmakers and component producers, even if some changes and customizations were necessary, due to the different social and cultural contexts. The maximum constraint of this model was mostly smoothing of production (point 6), which, in a highly variable car market, is often difficult to apply. Even in Japan at the end of this period, the economic crisis would heat the industrial compartment, causing it to react brilliantly in the next decade.

F. ***1995–2005 Market globalization and the integration of production processes on world scale***

Some important events such as the creation of Free Exchange Areas (European Community, NAFTA, ASEAN…), the reunification of Germany after demolition of the Berlin Wall, the re-opening of Eastern European markets, led "carmakers" and their suppliers to build important new plants in Eastern Europe (Germany, Poland), Turkey, Brazil and the Far East of Asia.

Globalized markets, the consequence of free trade policies, led to major companies giving serious review to their industrial strategies. The concept of a "world-car" and the strategy of producing and managing the supply of components on a world scale was introduced (*global sourcing*). Several industrial integrations and *joint ventures* were signed for the development of new products in "-makership".

For product development policies, brands tended to increase the number of specialized products by introducing new multipurpose vehicles (MPV, SUV, LCV, …), using standardized solutions for engines and space frames. Another distinctive phenomenon was the increase of diesel engines by direct injection, the so-called "common rail-multi jet" (solution designed by FIAT, developed industrially by Bosch, and soon applied by the greater part of European and Japanese carmakers, often with some technical improvements).

Introduction of new technology for the assembly of material allowed for lighter vehicle structures, and improved performance, fuel consumption and safety. In this area, the best performers were Audi and Ferrari for aluminium space frame solutions and Volkswagen for high resistance steels and the application of laser junction welding techniques.

Innovations in vehicle structure, and new generation diesel engine development, together with research for alternative energy supply solutions, led to a significant increase in economic expenses in R&D, as compared to the previous decade.

In the meantime, production capacities in the major industrial areas (North America and Japan, Western Europe and South Korea) were so large, the market was flooded. As a consequence, in the latter half of the '90s, the competitive scenario became critical, and most carmakers without a strongly established presence in every market found themselves in a financial crisis and were obliged to aggregate to bigger industrial groups.

Several concentrations and alliances between companies formed in this period would last into the next generation of important industrial assets. The principle ones were Renault-Nissan, Daewoo's acquisition by GM, Seat and Skoda's acquisition by Wolkswagen, the co-makership alliance between Fiat and PSA for light commercial vehicles and that between PSA and Ford for a new generation of diesel engines.

Conversely, other important "liaisons" between companies would be reconsidered as a consequence of unsuccessful results: the GM-Fiat alliance was changed from a total automotive collaboration to the single production of Diesel-Multijet engines, while the Daimler-Chrysler joint effort was simplified to the selling of Chrysler on the Daimler Benz side.

For other industrial compartments, such as heavy and medium commercial vehicles, or agricultural and construction equipment, similar alliances had already taken place in the past. For instance, IVECO, with the acquisition of the Fiat/OM, Magirus, Unique, and Pegasus brands, was one of the most important examples. At the same time, CNH, which joined Fiat New Holland and Case, also became a worldwide company with its own industrial compartment.

Strategic goals of these industrial integrations were:

- to reach important synergies from an economic point of view by using R&D on common platform and new investment;
- to increase contractual power in material, component and equipment acquisition;
- to increase market shares on worldwide markets by widening the number of models offered and leaving exclusive missions to single brands.

To reach these targets, the big carmakers followed the strategy of empowering R&D capacities and the distinctive features of single brands, at the same time assuring effectiveness in cooperation between the participant companies. Modern digital technologies for communication made data exchange and logistic flows easier on a world scale.

Even component and equipment suppliers followed the same strategies, assuming that "lean production" and total quality policies were the only way to be competitive in the challenge determined by the *carmakers*.

By the last period, Toyota and BMW had achieved major success, with market shares increasing and economic results constantly in the positive, without the integration of any other company and solely through investment in product innovation.

In spite of a scaling down in economies, interest in special vehicles for elite customers did not decrease. Some important brands used new technologies to develop original styles on a small scale and prestigious content for special cars: Porsche and Ferrari were the best brands for *super cars,* while other interesting sports cars were launched by BMW and Alfa Romeo.

G. ***2005–Today International Crises and New Worldwide Scenarios in the Automotive Industry***

The overall scenario in the automotive industry at this time was directly related to a big economic crisis involving all carmakers at a worldwide level. Overproduction came head to head with a contraction in the car market. This situation led some carmakers to the edge of bankruptcy, all over the world but especially in North America. In 2009, the US Government had to intervene to save jobs and offered the chance of a controlled administration, providing temporary funding to encourage new investments. The big three American carmakers, GM, Ford and Chrysler, were put under controlled administration until a new industrial scenario could be conceived.

Ford and General Motors, the two bigger of the three sisters, could count on more robust industrial organizations, and so, through a deep review of their industrial and commercial strategies, were able to refresh their positions and ensure the US government of a solid recovery plan.

Chrysler was coming out of a failed joint venture with Daimler and was incapable of reorganizing by itself. The American Government, after analysis of assorted possibilities, sponsored an alliance with FIAT that, in terms of product range and powertrain technologies, could offer the right guarantees of success. In only two years, industrial cooperation between the companies led to a complete repayment of the government debt, and by the end of the summer of 2011, FIAT owned the majority of the stakeholders, taking Chrysler out of bankruptcy.

In this scenario, new joint ventures, like FIAT and Chrysler, give new structure to the role of the carmaker, while solutions for a more environmentally sustainable vehicle are required to solve the car market crisis definitively and give new opportunities for refreshing the entire automotive industry, a scenario in which all remaining major carmakers attempt to move forward as fast as they can, looking also to the evolution of social and political situations in the big industrial areas. In such an industrial world, the survival of a carmaker is linked more than ever to the possibility of sharing costs through alliances, gaining the right economy of scale, and trying to minimize all cost factors linked to industrialization.

At the end of this historical overview, we must point out that the classification of different periods has been used only for didactics, while, in practice, each period is linked to the next by a continuous line and without a net separation.

Actual individual periods are still difficult to determine from the point of view of the automotive industry. In particular, European and North American carmakers have been reduced and have optimized their production sites. It is quite probable that the ratio between offering and market request will smooth out among the major producers in Europe, North America and Japan.

In the medium-long term period, *carmakers* and component producers who are successful in overcoming these difficulties will then have to deal with a new scenario of deep change, as foreseen by that most esteemed international agency as occurring between 2015 and 2020, due to certain macro-phenomena already happening:

- growth in the worldwide market in certain specific areas characterized by low manpower cost and a big potential market (China, India…)
- increase in the cost of petrol and natural gas (always critical resources in the global geopolitical situation);
- adoption of new *energy-saving* and *anti-pollution* policies (particularly severe), even in North America, beginning in the state of California and then spreading out to the other states;
- vehicle access to urban centres and commercial vehicle limitation on highways of high traffic density in Europe.

To be less tied to petrol, engines must be adapted to alternative fuels, derived from natural gas or from agricultural resources. Recently introduced hybrid powertrain systems, especially those by the Toyota Group, are very interesting from a performance and consumption standpoint in urban traffic, even if production costs are much too high to allow diffusion on a large scale. It has been estimated that, from 2020 on, electric powertrain systems equipped with fuel cells will be diffused, especially in urban centres in which anti-pollution restrictions are adopted.

On the other side, up to now major carmakers have made investments to develop engines fueled by diesel, methane and petrol: multijet systems are applied by using filters on exhaust. In this way, performance is improved and dangerous emissions are reduced at the same time. This will lead to some additional costs, but less so than for other new innovative solutions mentioned above.

Another strategy will be one for reducing the weight of vehicles, maintaining the same features in loading capacity, internal comfort and ergonomics, but at the same time improving active and passive safety.

Production systems will be progressively changed in relation to product innovations and the reduction of energy consumption on a CO2 plant emission level.

For all these reasons, to satisfy the demand for constant worldwide mobility, major vehicle producers and their suppliers (component and production systems producers) must cooperate with each other and with institutions to establish industrially feasible restrictions. Only if the afore-mentioned technological progress occurs will it be possible for an intensification of car traffic and commercial transport of vehicles. At the same time, public services dedicated to mobility must change in parallel, becoming systems of active information for vehicles.

Table 1.1 Global light vehicle sales trend—Units × 10^6 (Automotive News source)

Market	2011	Forecast 2014	Forecast 2018
Europe	18.4	19.5	21.8
North America	15.3	18.4	19.3
South America	5.5	6.5	8.4
India	3.0	4.1	5.8
China	17.6	23.2	29.2
Japan	4.1	4.6	4.4
World	75.6	88.9	103.2

Finally, to better understand the trend in automotive production activities on a world scale, we will examine the trend in global light vehicle sales (including cars, and medium and light commercial vehicles), divided into macro-geographical and economic areas. Table 1.1 resumes forecast data related to the periods 2014/2018 with a benchmark for 2011; the source is "Automotive News" 2012.

The biggest growth interests are in the Asian market, due to economic expansion in China and India with, respectively, 66 and 94 % market increases, but even countries in Eastern Europe (including Turkey) and South America (65 % represented by Brazil) are involved in this trend.

Conversely, countries in major industrial areas (Western Europe, partially balanced by Eastern European growth, North America and Japan) are interested in slow growth or, in the case of Japan, in decreasing volume as well. In these areas, industrial development will essentially be focused on quality, involving aspects such as vehicle class, powertrain systems, consumption/performances ratio, comfort and safety.

Globally, a 36 % increase in moving vehicles is foreseen by 2018, in spite of 2011, with a consequent growth of business in the automotive industry, even if new technological solutions become more and more necessary for solving environmental matters and energy source availability.

Concerning the geographical distribution of production sites, it is foreseen that by 2015 more than 40 % of the total sites will be represented by emerging countries outside of the current triad (Western Europe, North America and Japan). For these reasons, commercial trade to and from triad areas will be influenced by the following phenomena:

- increased export of products, systems and services with high technology content towards extra-triad areas;
- increased import of "labour intensive" standard products from Far East Asia and rough material from Russia and certain countries in South America and Africa.

As seen in industrial history, every time such a change occurs, there are major consequences in regard to industrial assets. Companies suddenly have to alter their strategies to guarantee their shares on the international market and set out important financial resources for introducing innovation and finding new opportunity for business.

At the end of this historical evolution, we can consider what follows:

- **Change is a slow process that may very well become necessary, so, for a company's survival and success, it is important that the signals of imminent change be recognized so that a company can react quickly.**
- **Production systems improve gradually, due to the pre-existing industrial systems and occupational agreements; nevertheless, company strategies must constantly be focused on product/market targets, with a frequent assessment of one's closest competitors.**
- **Industrial alliances are often a necessary step in approaching a new market, while product innovation and new technologies are necessary as well; cooperation must be considered to maximize benefit and convenience in expanding product/market areas.**

In the following, we will deal with criteria applied for the establishment and management of production processes much more deeply related to the development of competitiveness of companies and their environmental and energy-related sustainability.

As we will see in the next chapters, the most critical factor for the success of industrial projects is the ability to develop "lean" systems (easy to convert and to maintain), fully responding to "quality and innovation targets" for products.

1.2 Strategic Planning of Production Activities

Industrial strategies for development derive from studies of the evolution of market and customer demand. They have to consider the existing technology to make it easier to determine opportunities for innovation of the product and production processes. To identify priorities for research and investment plans, it is fundamental to evaluate the existing and potential industrial capacity of the company realistically, having in mind its weak and strong points, to direct efforts toward key success factors.

Figure 1.2 shows the three typical aspects involved in strategic planning. The final target to be followed in cooperation with the different departments of a company is the "*value generation*".

Company resources involved in the production process and productivity key management indicators

Our focus is now on production processes, for which the major part of manufacturing company resources (manpower, investments on systems and work in

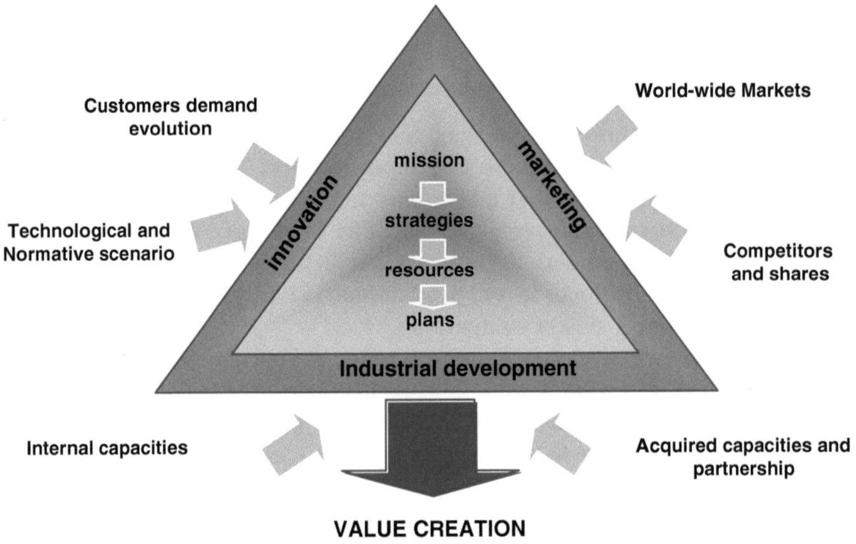

Fig. 1.2 Activities for strategic planning process

process, materials and energy) are normally employed. The typical key management indicators used to evaluate the effectiveness of the employment of resources, in terms of output/input ratio, are as follows:

1	Manpower productivity	Quantity of transformed product/quantity of manpower or quantity of man-hours dedicated
2	Production systems productivity	Quantity of transformed product/nr of days or hours used for producing
3	Direct material utilization degree	Net direct material requirement based on design elements (final product)/gross requirement based on transformation process
4	Energy and auxiliary materials productivity	Quantity of transformed product/specific energy and auxiliary materials used in transformation processes
5	Investments productivity	Production capacity used during product industrial life cycle/product specific investments

Increasing these productivity factors helps to reduce the cost of production and get the gross profit margin in positive numbers considering the same sales value/price of the products.

For this reason, in the following chapters we will be dealing with methodologies for analysis and improvement of such productivity indexes. We should also say that improvement plans must extend to the entire "supply chain", in consideration of an integrated production process.

Other strategic key success factors

The above Key Management Indicators are directly connected to product competitiveness, while sales success and company results are also dependent on the following factors:

- *Product quality and innovation*
  Value/cost ratio determines product profitability, which depends mostly on the innovative capacity of the company and on the quality level perceived by customers.
  This argument is considered in Chap. 8, dedicated to "continuous improvement" management.
- *Customer service level*
  This represents the capacity to respond in terms of timing, quality and quantity to the fulfilment of obligations engaged with final and intermediate customers. Methodologies for achieving a level of service towards customers are addressed in Chap. 6, dedicated to Management of Logistic Processes.
  It is very important to underline that the capacity to fulfil the product range in the required mixed model, with short process lead time, is strictly related to the strategies adopted for engineering of "Production Systems", as we will see in Sect. 1.4.
- "*Time to market*"
  This is the period of time necessary for the technical and industrial development of a product, starting from the definition of the "concept" or "style model" up to the commercial launch. The capacity to launch a new interesting product faster than one's competitors is directly linked to success in sales and the company's resulting economics.
  This argument is considered in Chap. 2, dealing with new product development and engineering.
- *Product and company brand image*
  This fundamental aspect will be not approached in this text because it is an argument for marketing strategies. What is important to emphasize here is the fact that product and company brand image depend not only on communication strategies but especially on the perception by customers of a particular level of quality. Brand image is sustained by product innovation that does not sacrifice the essential marks of its origins, considered important for brand distinction.

Industrial development and life cycle of a product

Figure 1.3 shows the macro phases that feature in the industrial life and development cycle of a product:

The development of a new product requires preliminary research activities for establishing performance and qualitative targets born out of the stated mission of the brand or company. Parallel to this, a qualitative and quantitative analysis of market requests is run to define sales targets, as well as their relation to the best of one's competitors. In this way, the level of production required can be planned on a sufficiently wide temporal horizon.

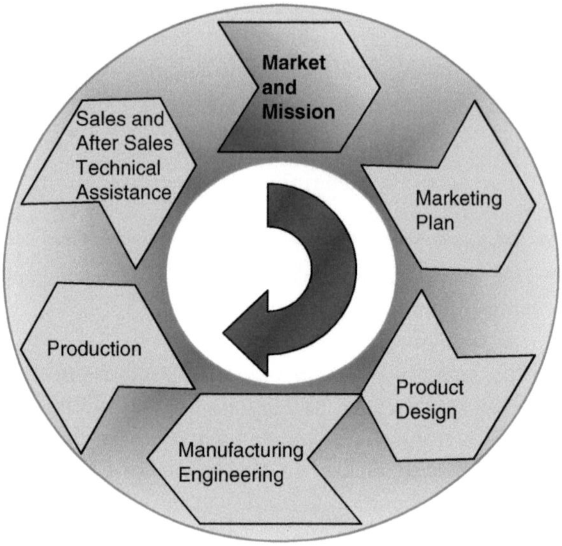

Fig. 1.3 Product industrial cycle life

After this phase, executive project and manufacturing engineering activities follow. The *development cycle of a new product* ends with the start of production and commercial launch in the market.

The industrial life cycle of a product begins with start of production and ends with completed sales in the market, even if the company has to provide technical assistance and physical maintenance to customers subsequent to the sales. Another parameter is the *technical life cycle*, which depends on the product's functional duration (the actual use) and on technical and economic considerations.

In the following chapters, we will take a wide approach to matters related to new product industrialization and the management of their production systems, referring from a general point of view to their *industrial development and life cycle.*

1.3 Process Integration and "Make or Buy" Decision-Making

A global industrial process includes everything from the final productions managed directly by the "carmakers" to the intermediate productions managed by first level suppliers and their sub-contractors, in other words, the so-called "supply chain".

In defining strategies for production activities, each of the companies involved in the global industrial process has to establish which parts and modules must be manufactured in their own plants, which have to be realized in "co-makership" with other companies, by joint ventures and strategic industrial cooperation, and

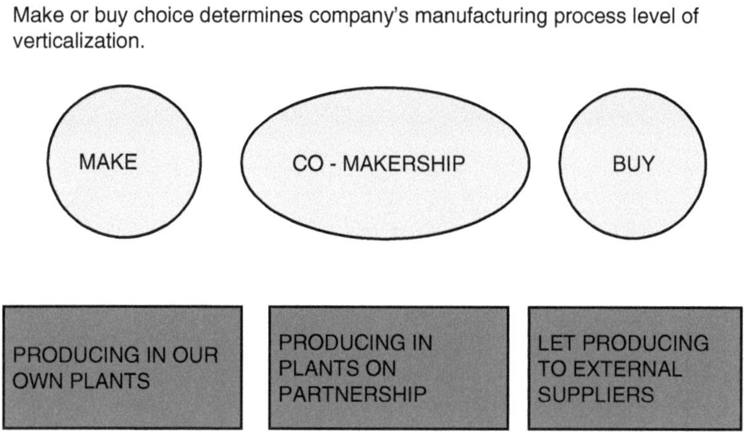

Fig. 1.4 Make or buy policies

which must be bought from external suppliers, in accordance with the company's top management policies (Fig. 1.4).

For big companies, the "make or buy" decision-making process involves the following departments:

- Product Development Planning
- Product and Process Engineering
- Manufacturing
- Purchasing and Supply Chain Management
- Finance
- Human Resources

Decisions are made by the appropriate executives responding to a Steering Committee, which defines strategic policies for the assurance of "governance" in investments and purchases.

"Make-or-buy" criteria for whole types of systems and components require a very complex process of evaluation. The most important items in this process are the following:

1.3.1 Items (m) for the Decision to "Make"

m1 – Availability of development capacities inside the company, at least equivalent to those offered by external suppliers

m2 – Better protection of company's know-how, in relation to products/processes considered part of the "core business"

m3 – Need for the integration of internal processes by logistic flows and the improvement of the quality level of the final product

m4 – Established invalidity of alternatives for the purchasing of components or services
m5 – Necessity of utilizing existing productive capacities inside the company

1.3.2 Items (b) for the Decision to "Buy"

b1 – No availability of specific "know" for developing the specific component or service in a competitive way
b2 – Availability of reliable suppliers for the components or services required, in relation to the company's targets
b3 – Availability of alternative supply sources in the geographical areas of strategic interest for the company
b4 – Opportunity to simplify internal production processes and focus production resources (manpower and financial resources) on "core-business"
b5 – Opportunity to reduce company-owned investment by using or incrementing supplier's available productive capacities, allowing them to assume the risk of volume dependent on market trends

1.3.3 Items (c) for the Decision for "Co-Makership"

c1 – New necessities for the development of production capacities but not enough economic scale to proceed autonomously
c2 – Partners have common or complementary interest in developing the new solutions required to guarantee the industrial mission in a synergic and competitive way
c3 – Partners have complementary or synergic technological capacities at their disposal so that product/process innovation can be sped up without incurring economic investment

The above decision-making criteria establish the guide line for *"make or buy"* choices and/or industrial cooperation and, as a consequence, operative instructions are assumed.

At any rate, specific analysis may be required for new industrial initiatives, comparing economic data of alternative solutions by considering:

- Variable production costs (for direct material and for transformation process);
- Fixed costs consequent to company's R&D department;

- Investments for internal manufacturing or external purchasing (suppliers' contribution);
- Other eventual costs derived from fixed costs.

"In-sourcing" or "out-sourcing" policies involve not only product components but research and development services, equipment construction, maintenance, logistics, information technologies and other accessory services. The criteria followed are almost the same as above and must take into consideration economic advantages obtainable over long periods (from 2 to 4 years), involving specialized companies operating on a large market scale (*system and service providers).*

"Make-or-buy" choices influence a company's structure and assets, capital invested and economic profitability. For these reasons, an accurate cost/benefit analysis for a medium/long term period is needed, making clear that incoherent decisions could highly compromise the level of quality and competitiveness of the final product.

The "verticalization degree" **of a company's production activities is represented by this ratio: total cost of internal manufacturing processes/total cost of sale products plant free.**

Beginning in the second half of the 1980s, carmaker strategies have been oriented towards progressively reducing the verticalization degree. Conversely, companies have widened the range of products offered, increasing their presence on international markets.

Production activities verticalization in plants manufacturing cars and light commercial vehicles has a range of a 25 % minimum up to a 30 % maximum, including the manufacturing of powertrain systems (autonomous or in *co makership*) and excluding components produced by companies that are part of the same industrial group but operate independently on the market. Industrial vehicle plants normally have a minor verticalization degree.

Actually, in Europe, the more vertical carmaker plants are those of Volkswagen, Mercedes, P.S.A. and Auto Vaz, while the less vertical plants belong to BMW, Volvo, Saab, Jaguar and Porsche. North American and Japanese plants generally have a minor verticalization degree in spite of the example of those in Europe.

1.4 Manufacturing Systems Set-Up and Location Criteria

Referring to the carmakers' and component producers' "best practices", Fig. 1.5 represents how the decision-making and management process of the industrial initiatives for new product development, or production capacity empowerment, work.

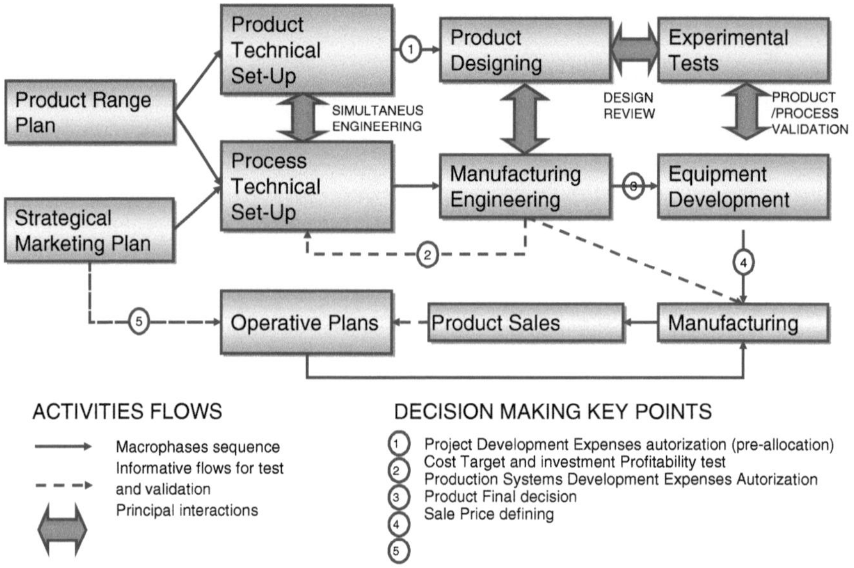

Fig. 1.5 Product initiatives setting and controlling process

1.4.1 Equipment Level and the Defining of Technological Solutions

In relation to "making" processes, we will deal with the applied criteria in terms of strategic setting.

For each product family, the necessary production levels (maximum quantity for a year and a day) and the level of flexibility for the production mix-models are defined, based on information coming from the product range plan and sales forecasts.

The "product range plan" refers initially to basic models, powertrain and space frame families. Information is then refined into specific vehicle or space frame versions and, finally, the optional contents are considered, and a commitment is made to those important for the industrial structures (for example: automatic gears, integral powertrain systems…). Starting from the "product range plan", the "strategic component plan" is derived, considering the synergies for similar components and sub-systems, and referring to capital intensive modules, for what medium-long term planning is required. The "product range plan" reports the industrial life cycle of the models divided by engine versions, both the basics and then the derived versions, and it also indicates the date (year and month) of commercial launch for the new products in substitution of the old ones.

Equipment level definition (production capacity and the principal features of production systems) derives from the above "product range plan" and from the production capacity required. For this reason, marketing activities are very important; they give origin to the "strategic marketing plan", which refers to a long period of time (for the automotive industry, the period is normally ten years) in which market shares are indicated year by year and model by model.

In relation to the production capacities required and to the best economic solutions for the processes, a "manufacturing engineering plan" is derived, from which equipment and specific tools choices are made.

This argument will be addressed in Chap. 3. It is important to understand that technological equipment solutions depend on manufacturing scale and on the flexibility required by the production of mixed-models, based on the "strategic marketing plan".

To exemplify these concepts, the following "Productive Scale/Technological Solution Correlation Diagram", shown in Fig. 1.6, is relative to the thin sheet metal part printing process for bodies:

Quantities to be produced in a year and their relative medium quantities a day are reported on the "x" axle, on a logarithmic scale, while the manufacturing costs per unit are reported on the "y" axle, in relation to the single technological solutions taken into account and reported at the bottom of the page.

Each intersection between two curves defines the areas of better economic convenience.

Similar diagrams can be designed for other manufacturing technologies, for example:

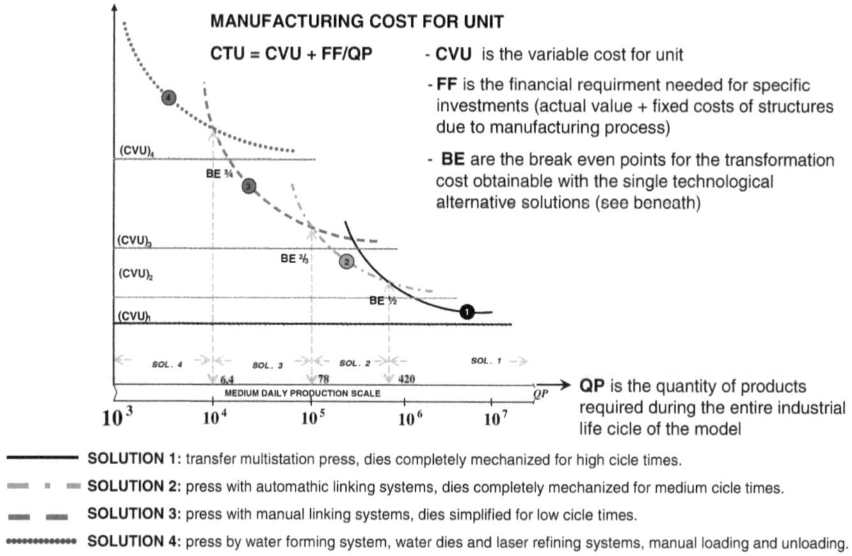

Fig. 1.6 Productive scale technological solution correlation diagram

- for final assembly operations, completely manual assembly stations for little series production are followed by automatic operation cells for medium series production, and then full automatic systems for big series production;
- for mechanical machining, by increase in production scale, "stand alone" machine tools are followed by multi-operational "machining centres" and then "transfer lines" for high cycle times;
- for foundry processes, by increase in production scale, casting manual stations are followed by automatic cells and then completely automatic lines for high cycle times.

In searching for the best technological solutions, in addition to considering the productive scale, it is also important to analyse specific features for each product, and identify common and distinctive aspects for each part, in relation to the transformation processes. According to the availability of flexible and convertible lines, considering also the production volumes and the mix delivered, the best and most effective solutions for high productivity and flexibility are chosen.

Criteria for choosing a location for manufacturing activities

In deciding the most convenient place for the location of production capacities, alternative solutions should be considered, based on the following questions:

(1) Is it convenient to locate the production activity in an existing production site, with an established level of technological know-how and manpower availability?
(2) Are there existing production sites with sufficient area and basic equipment to allow for a significant reduction in the initial investment as compared to starting from scratch?
(3) In the case of new solutions, where are the best conditions to be found in terms of scouting human resources and a supply of energy and material at the lowest industrial cost?
(4) Among available geographical locations, which is the more convenient from the point of view of material and component supply (also considering custom fees)?
(5) Which is the more convenient geographical solution for dispatching the final product to the markets?
(6) In case it is convenient to divide production activities between two or more sites, what solutions make for more effective operative flexibility?

Along with the above, it is also important to consider these other geographical considerations:

- Legal conditions and fiscal policies favourable for industrial development;
- Existing infrastructure for the delivery of goods, supply of water and energy, and process material exhaustion;
- Ease in recovering necessary manpower and the possibility of extending working times, for optimal use of production systems;

- Probability of obtaining quick authorization from the local legal entities for building and general equipment construction;
- Availability of potential funds for employment development in specific areas.

The location of manufacturing activities for the same product and planned volumes can also influence the automation level processes. In fact, the higher the manufacturing cost, the more convenient it is to parse out the automation level. Criteria for analysis are the same as shown above in the diagrams correlating volumes/technologies, considering the right cost parameters in relation to industrial site choice.

Based on the above items, it is clear that decisions concerning equipment level and manufacturing locations require a complex decision-making process that must be supported by specific economic investigations.

Finally, we resume the ultimate purpose for the strategic planning of setting technologies and the location of manufacturing activities:

- **setting production capacities according to medium/long term marketing plan**;
- **assuring necessary qualitative and quantitative levels for products, at minimum industrial cost, including direct material, its transformation and assembly, final tests and their dispatching**;
- **assuring delivering lead times and service levels useful for the commercial and after-market networks**;
- **restricting the company's financial resources, considering cooperation for "co-makership" and local public fund opportunities**.

1.5 Overview of Technologies for Materials Applied to the Construction of Vehicles

Here is a list of the principal technological processes applied to the production of specific mechanical parts and the assembly of powertrain systems, bodies and final vehicles:

1. <u>Foundry and League Treatments Processes</u>: cast iron and aluminium foundries by sand or semi-permanent castings. These processes require basic investments and are labour, mining and energy intensive. Furthermore, exhaustion and regeneration costs for the processing of residual materials are required.
2. <u>Massive Steel Parts Printing</u>: forging, hot and cold extrusion, combined cycles. These processes require great investment, are electric energy consumption intensive and use processed materials with high costs (rapid wear moulding). Their merit is the obtainment of parts that do not require other machining operations for over metal removal.

3 Composed Metal Dust Sintering: these processes require a special technology and significant basic investment. Same considerations as previous point.
4. Metal Removal Machining, Heat Treatments and Super Finishing: by special machine tools, machining centres, multi-station and multi-tools systems, and equipment for heat treatment and special antiware metals. These processes require initially high basic and then product-dedicated investment.
5. Thin Laminated Steel and Aluminium Parts Printing, by shearing machining, printing, refining, folding and coining under press machine. These processes require initially very high basic and then product-dedicated investment. The degree of construction specifications for mouldings and raw material usage influence economic convenience of the process and quality of the final product.
6. Plastic Parts Printing, by injection printing, injection-compression, extrusion and relative processes for coating and joining parts. These processes require initially high basic and then product-dedicated investment. Requirements of material features and construction highly influence the economic convenience of the process and quality of the final product.
7. Assembling and Joining of Steel Parts: by resistance welding, laser welding with or without metal amount carried over, MIG/TIG welding, cold welding, etc. These processes require initially high basic and then product-dedicated investment and are also labour intensive. The degree of technical features of tooling and equipment utilization influence the economic convenience of the process and quality of the final product.
8. Protection Coating and Painting Process for Bodies, Cabins and Space Frames: these processes require high basic investment, and are energy, process material and even labour intensive. The degree of the material's technical features and utilization for these resources highly influence the economic convenience of the process and the quality of the final product.
9. Mechanical Groups Assembly and On Vehicle Installation: these processes require high basic investments, and are energy, process material and even labour intensive. They also involve high logistic complexity and have a high impact on product availability and quality level from a customer point of view.
10. Modules and Vehicle Final Assembly: these processes require high basic investment and are labour intensive. They also involve high logistic complexity and have a high impact on product availability and quality level from a customer point of view.

Carmakers and mechanical parts producers must exert tight control over these ten technological areas mentioned above, directly, when processes are developed inside the company, and indirectly, when they are the work of external suppliers, but still controlled by the carmakers. These technological areas are strictly related to the "supply chain" phases and require specific investment for product design, normally provided by the purchaser.

Furthermore, automotive production also requires other important technological contributions, typically developed autonomously by specialized suppliers, which are independent from carmakers. These technologies are connected to the following elements or vehicle sub-systems:

(a) *Powertrain Systems*

- High Specialization Standard Mechanical Components (pistons, valves, oil pumps, turbo chargers…)
- Engine Supply Functional Modules
- Exhaustion and Silencing Functional Systems
- Engine Thermal Systems
- Powertrain Transmission Functional Modules and Special Components
- Fluidic Tight Capacity Plastic Elements
- Engine Electrical System Components
- Powertrain Electronic Control Systems

(b) *Vehicle General Assembling Systems*

- On Board Instruments and Info-Systems
- Air Conditioning Systems
- Lighting and Vision Functional Modules
- Seat Modules
- Door Opening and Closing Leverages
- Safety Systems for Body Cell (air bag, safety belts…)
- Fluidic Tight Capacity Plastic Elements

(c) *Vehicle General Systems*

- Tires and Wheel Rims
- Breaking System Functional Modules and Special Components
- Suspension Functional Modules and Special Components
- Steering Wheel Leverage Functional Modules and Special Components
- Fuel Tank and Pump Functional Modules
- Electric Power Supply Modules
- Electric Wiring and Connections
- Oleo Dynamic and Pneumatic Equipment Elements
- Acoustic and Thermal Insulation Element
- Vehicle Electronic Control Systems

Those elements and systems are based on "evolved technological solutions", the "know-how" of which generally belongs to specialized companies that produce components; these companies are able to supply all carmakers on a large market scale, being the owners of autonomous R&D departments. Anyway, a strong cooperation between those companies and the carmakers is necessary in case of the development of technical projects and experimentation and homologation on vehicle phases.

It is relevant to observe that electrical and electronic contents included on vehicles have grown exponentially over the last decade, a trend that will probably only increase in the future. At the same time, technological evolution of microchips will allow for a reduction in impact on cost and availability of products.

To complete the vision of the "supply chain" involved in the automotive manufacturing processes, we also have to talk about basic material transformation technologies that take part in the structural composition of the vehicle:

- steel casting, lamination, extrusion and refining
- aluminium league casting, lamination, extrusion and refining
- magnesium league casting
- simple and strengthened polymer material technologies
- glass parts lamination, forging and cutting
- body protection and painting material technologies
- lubrication technologies
- electric conductor and optic fibre technologies
- electrical accumulator and catalyst converter special material technologies
- lighting and reflecting systems technologies.

Even technologies typical of these commodities require highly specialized know-how, reserved for suppliers. In the design phase, suppliers are called to define application and transformation characteristics of materials.

In Chap. 7, "Purchasing and Collaboration", we will deal with supplying policies and the control of component and raw material cost. According to modern purchasing policies, supplier's selections are defined in regard to the above technological classifications.

The total amount of items contributes to the increasing complexity of technologies applied in automotive industries. In setting and managing industrial collaborations, it is necessary to have a systemic ownership, so that it is possible to apply the different specialties in relation to product/market targets to generate value for the companies.

The following composition diagram shows the medium incidence of the weight of the principal raw material used in carmaking. They are indicative values, coming from an analysis on models with higher volumes in the European Community in the period 2001/2003, with reference to the trend of the next seven/eight years.

It is possible to note that metal and iron leagues, in spite of several reductions in the past, represent more than 75 % of the total weight of the vehicle, even if in the future their incidence will be reduced as a consequence of the employment of high resistance steels and light leagues (aluminium and magnesium) and, in addition, plastic reinforced materials. Another trend will be the growth of ceramics and metal matrix composite materials, useful for improving the effectiveness of powertrain and braking systems and reducing pollution emissions. With the introduction of electric and hybrid traction in the future, we will have a higher impact of copper, silicon and special metals used for catalicity converters and electric accumulators over a long term period (Fig. 1.7).

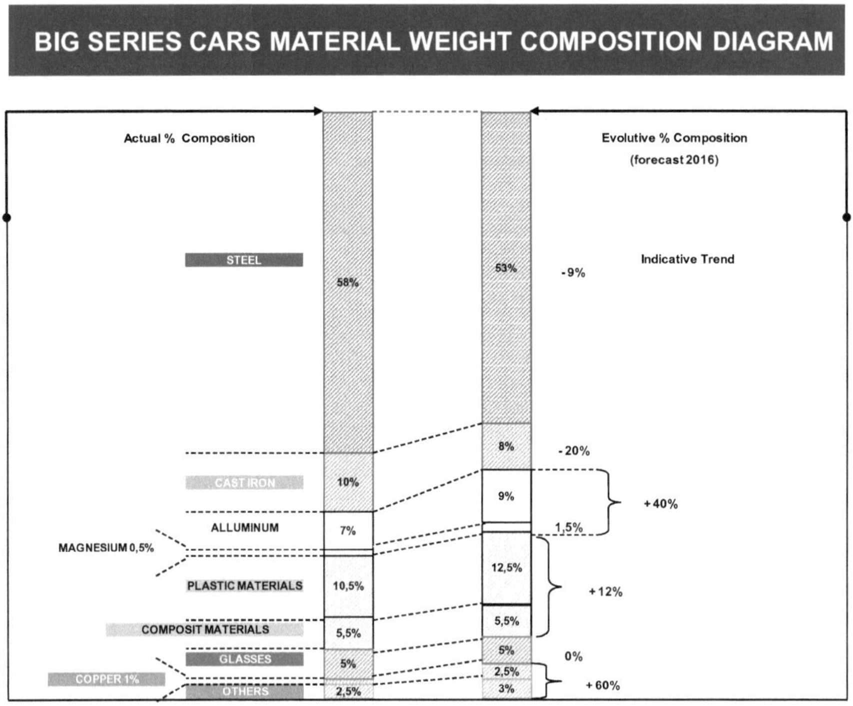

Fig. 1.7 Cars mass production material weight composition diagram

1.6 Sketches for Manufacturing Systems Adopted in Car Manufacturing

The following sketches show indicative diagrams of the plants where final general assembly of cars and powertrain systems takes place. Our representation is limited to block diagrams referring to the modern shop floor for big series production.

For the several types of technologies, principal features and system setting logistics of the process are resumed. It is also considered a productivity level indicator in relation to the economic scale. This does not imply that the maximum level of the scale must be reached, because other variables influence the activity level (high class car manufacturing for elite customers, industrial vehicle manufacturing…).

First of all, we describe the cars' final general assembly process, which is normally organized on four different, separate shop floors with autonomous general equipment in each one. These shop floors are interconnected to guarantee an integrated logistic flow, with minimal lead times and stocks (Fig. 1.8).

The above blocks include:

(a) printing of principal body sheet metal parts (press shop)
(b) assembling and finishing of raw bodies (body in white shop)

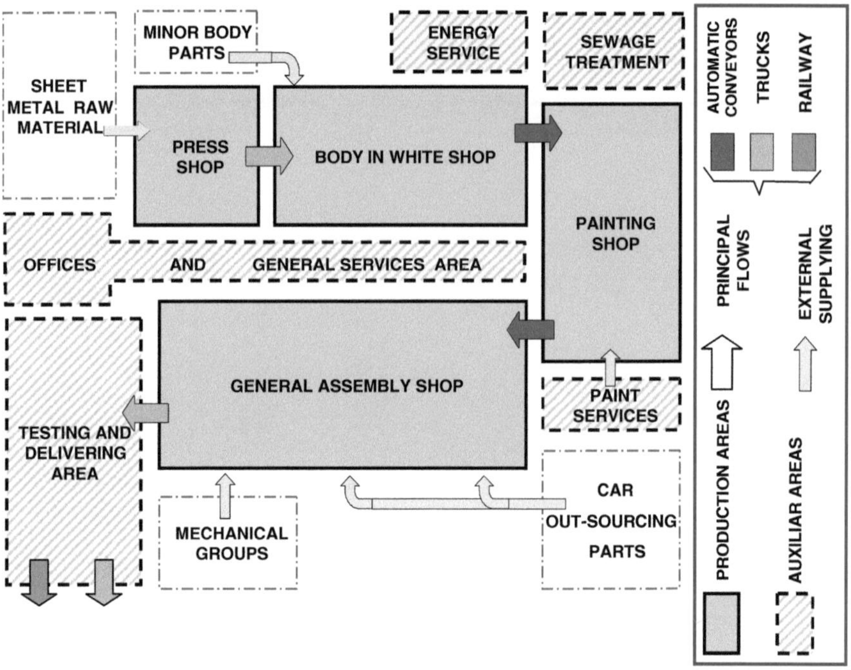

Fig. 1.8 General layout for a car assembly plant

(c) protective coating and painting of the bodies (painting shop)
(d) decking and final assembly of vehicles (general assembly shop)
(e) functional testing, sales lines and delivering (testing and delivering area).

In the following pages, the descriptions of the above process areas are described (Fig. 1.9).

- Biggest sheet metal part printing, used for body assembly, is done in "press centres" integrated or close to the final assembly plants.
- The printing cycle of the principal elements of the vehicle bodies includes more operations (from a minimum of four to a maximum of six), performed under mechanical or hydraulic presses up to a printing strength of 2000 tons for single presses and up to 10,000 tons for multi-station transfer presses, in sequenced and continuous cycles.
- Equipment and systems used are different depending on set production levels:

(a) Transfer multi-station mechanical presses with a very high cadence (over 15 cycles/min) fitting with a very high production rate (over 600 series/day);
(b) Traditional mechanical or new generation hydraulic presses with a medium cadence (between 10 and 15 cycles/min), interconnected by automatic systems, fitted for medium production rate (included between 60 and 600 series/day);

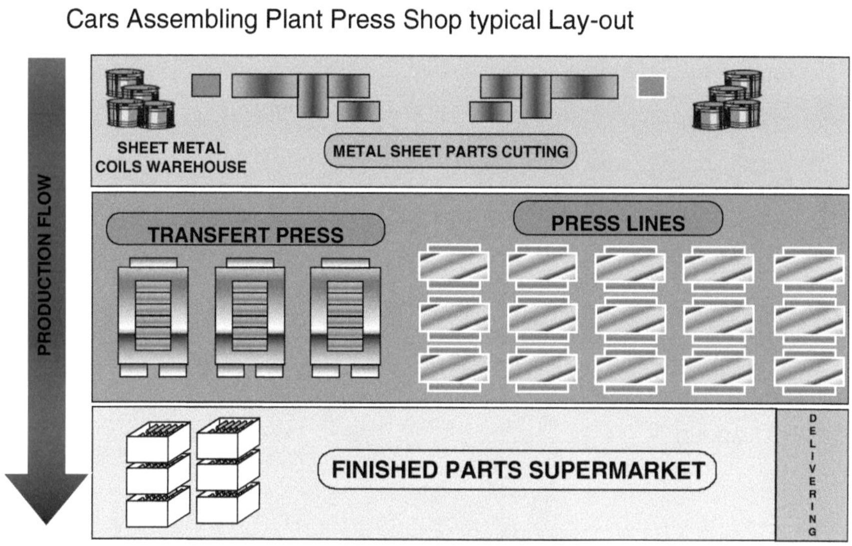

Fig. 1.9 Press shop layout

(c) Traditional hydraulic or water forming presses, with manual loading/unloading, with low cadence, fitted for low production rate (less than 60 series/day).

Major factors influencing manufacturing costs are: press and mould depreciations, maintenance costs and material handling. For these reasons, it is very important to reach and maintain a high level of OEE (Overall Equipment Efficiency) and a good utilization level (see Chap. 3) (Fig. 1.10).

Furthermore, to keep product costs low, usage of sheet metal must be optimized. This depends on cutting schemes and printing methods and technology ("sheet metal press" dimension). Overused metal parts are automatically conveyed in compacting equipment and transferred to where they can be reworked in metallurgic processes.

- The manufacturing engineering plan defines printing directions and pressing strengths necessary for the single operations.
- Mouldings are built in two principal twin parts (matrix and die), characterized by complex shake and hydraulic and pneumatic leverages. Matrix and die are connected in order to the fixed lower part and upper pressing part of the press machine.
- Production is organized by batch, stocking elements in specific containers, engineered to avoid risk of damage.
- Even basic investments (press machines) as well as specific ones (moulds) are very high; so it is very important to assure the overall effectiveness of the best production systems. For this purpose:

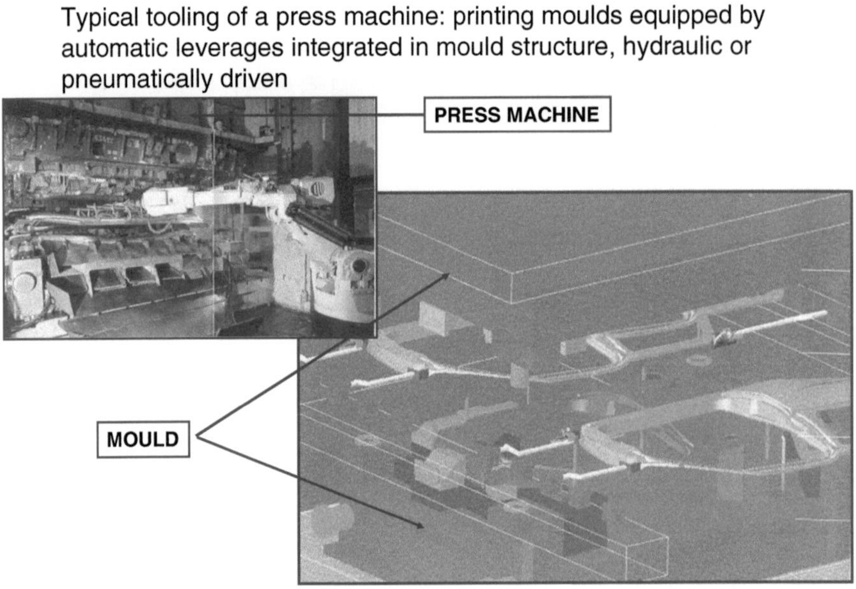

Fig. 1.10 Sheet metal parts printing process

- it is necessary to provide efficient preventive maintenance of moulds and press machine parts and components;
- fast systems for the changeover of moulds must be used to assure short "set-up times"
- (typically less than 15 min, from stoppage to restart of production).
- Geometric quality is statistically controlled through programmable measuring machines.
- Surface quality is normally checked through "on-line" visual controls, based on customer perception (Fig. 1.11).

- Following the actual guidelines, for high or medium volume body in white welding and assembly, hard automation (robot-intensive-oriented) and flexible production systems are used.
- Assembling tools are specific to each part of each product and are used to assure the "process capability" level necessary for geometric and style characteristics.
- Systems flexibility/convertibility is achieved through the rapid interchange of specific tools, so that it is easy to set the mix model level on the same equipment or line.

Major factors influencing manufacturing cost are: specific equipment and tool depreciation, maintenance and material handling costs. Even in this area it is very important to achieve and maintain a high level of overall equipment efficiency and a good utilization level (see Chap. 3) (Fig. 1.12).

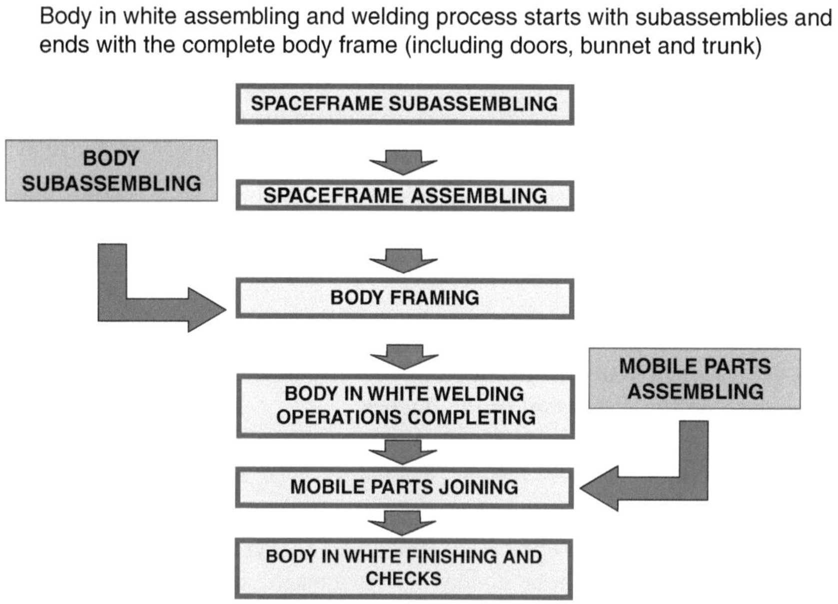

Fig. 1.11 Body in white welding and assembling flow

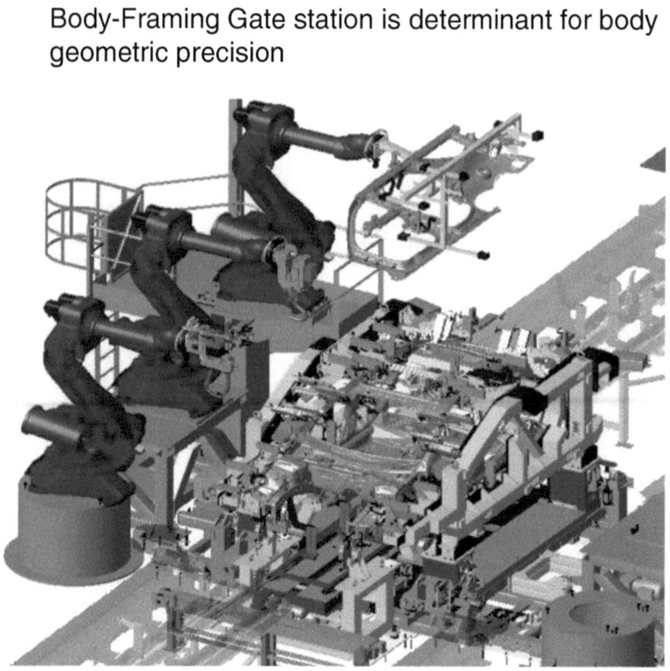

Fig. 1.12 Body in white framing gate station

- Joints are mostly formed through resistance welding performed by pneumatic and electro-mechanical welding guns. For aluminium structures, riveting and TIG (Tungsten Inert Gas) welding technologies are used. Recently, new joining technologies such as laser welding, with or without the material amount taking over, the riveting of hybrid joints, and structural gluing have been implemented.
  - In the completion line of the body in white welding operation, mobile parts are joined to the body and rightly adjusted together with plastic parts that must be painted on the line with the body.

- **For this technology, the manufacturing volume/investment ratio shows its best values when standard production capacity is set at 60/80 bodies/hour, depending on the dimension of the models**.
- For "special bodies", small and medium series, the technological layout is set by working cells, with the manual loading of parts and only partial robot welding (Fig. 1.13).

The body in white painting process is structured in sequenced phases, according to a continuous flow that can be divided into two macro areas:

(a) **Pre-painting treatments**, including washing, degreasing and phosphate treatment, which activates the sticking of the coating on metal, anti-corrosion coating applied by immersion of the body in an electro-chemical bath, application of polymeric materials for soundproofing and sealing.

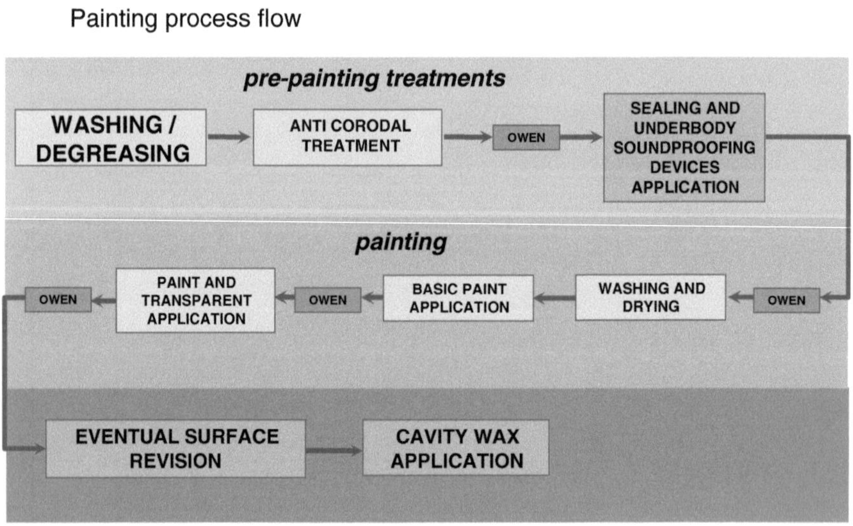

Fig. 1.13 Body in white painting process

(b) **Painting**, applied by automatic robots with an electrostatic spray, in conditioned cabins with a high level of air change, controlled temperature and humidity, so as to assure the elimination of the "over spray" phenomenon and fast evaporation of solvent or basic water coating.

After each phase, polymerization is completed by layer exsiccation, obtained by continuous flow ovens, and followed by an accurate washing with deionised water.

Focal points for the correct management of the above processes are:

- material usage control (high quality and expensive ones);
- fluidic and filtering equipment preventive maintenance;
- a rigorous clearing of the above equipment;
- painting and spray systems logistic management;
- quality control by process parameters and quality of manual job leading.
- The gaseous, liquid and solid process materials rejected are also relevant, and must be treated before exhaustion according to strict legal guidelines.

For this technology, the manufacturing volume/investment ratio shows its best values when standard production capacity is set at 60/80 bodies/hour for part a) of the process and at 30/40 bodies/hour for part b). It is thus normal for the same pre-painting treatment equipment unit to supply enough work for two painting equipment units.

Major factors influencing manufacturing cost are: specific direct material and energy consumption, manpower necessary to manage the process and equipment maintenance costs (Fig. 1.14).

For continuous flow in the general assembly of cars and commercial vehicles, or stop and go, lines with interconnected stations are used, managing fixed sequences of production. Lines can be divided into the following principal assembly sub-processes:

(a) **Pre-Decking Assembly**

It starts with doors disassembled from the painted body. The doors are removed to facilitate assembly of the operation inside the body and, once the dressing is completed, they are handled by automatic conveyors to the final assembly area; the subsequent operations are fluidic and electrical installations, cockpit, instruments, steering leverage and fuel tank assembly on the vehicle.

The painted body is moved to assembly stations through a double rail chain conveyor or self-moving conveyor, equipped with a hanging hook able to rotate across the translation axle to facilitate the best ergonomic condition for underbody operations.

Completed cockpit module and instruments and door modules are prepared in dedicated stations to the side of the principal line, and conveyed directly to the point of installation on the vehicle, according to the production sequence scheduled through programmed systems.

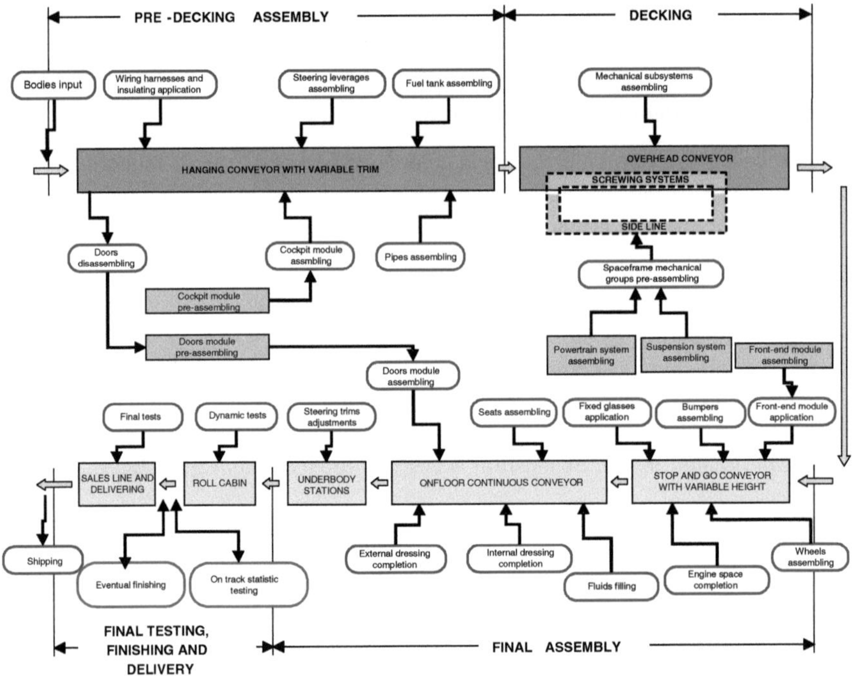

Fig. 1.14 General assembly process flow

(b) **Mechanical Assembly and Decking**
This important assembly phase of the process is made on line immediately after the previous one and includes the on-body assembly of the following mechanical groups: powertrain system, transmission groups, suspension systems, and exhaustion pipes. These groups are prepared in the "mechanical assembly" area and transferred to the "decking" area by synchronic conveyor systems. The joining to the body is normally performed by multi-head automatic screwing systems, with controlled fastening torque.

(c) **Final Assembly**
This phase is also connected to the previous one, but the same "decking" line could feed two parallel "final assembly" lines, in case it is necessary to divide production flow for different models with the same space frame. Each "final assembly" line is divided into three sequential parts.
In the *first part*, preassembled bodies coming from the decking area are put on special supports to be transferred by a stop and go system to the next station. In the modern plants, the bodies are arranged across the line axle in this phase to facilitate "front-end" module assembly and to keep electric battery, air filters and engine connections operations ergonomic. Fixed glasses are also assembled by automatic systems in this part. For the subsequent operations, such as wheel assembly, the body is again arranged on the line axle.

In the *second part*, the wheeled vehicle leans on a continuous moving conveyor belt on the floor line; seats, garnishes, door modules (pre-assembled to the side of the line and conveyed to the same original body), lighting systems and the rest of the vehicle parts are assembled.

In the *third part*, work is done on the underbody, adjusting the vehicle suspension and steering wheel regulation.

(d) **Final Testing, Finishing and Delivering**

This phase includes a test of the quality conformity control of electrical connections and functional dynamics (performed through a roll test while on-line), in preparation for the final testing. Subsequent refining operations are made "off-line" and can be considered to be a critical phase of the process

Truck or road testing is performed on 100 % of high performance cars and vehicles with a very high technical complexity, while, for cars and commercial vehicles of high production volume, only a set percentage are tested, following a statistical model. The latter assumes that a solid level of reliability has been reached through the entire production process, having passed the start-up phase of production.

Finally, vehicles are refined with accuracy and delivered to shipping, after the application of protections for avoiding damage during transportation and stocking phases.

Joining operations, screwing of mechanical groups and application of fixed glasses are generally performed by automatic systems, while the most labour intensive operations, such as application of the cockpit module, steering leverages, seats, and mobile parts, are done with the aid of partners to facilitate operations from an ergonomic point of view.

For manpower and investment productivity, the most effectively operative "decking" areas will average 60/80 vehicles/hour, while the "final assembly" areas will average 30/40 vehicles/hour. This means that the same "decking" line can supply two different "final assembly" lines. In this way, flows can be divided facilitating the supply of specific components and work loading for different models with the same space frame platform.

In this area of high logistic and organizational complexity, it is very important to use the modern product/process information technology known as "digital factory".

According to "lean production" principles, carmakers tend to organize body subsystems in preassembled modules to be tested separately (doors complete with moving windows, steering wheels with air-bag and electronic controls, cockpit modules fitted with air bag and instruments, front modules with thermal systems…). These modules are supplied "just-in-time" to the final assembly line. The modules have been tested beforehand, so that functional nonconformities that require "off-line" reworking operations decrease.

For components influencing vehicle dynamic safety, it is mandatory that the "traceability" of supplied batches be well-managed, assuring responsibility on

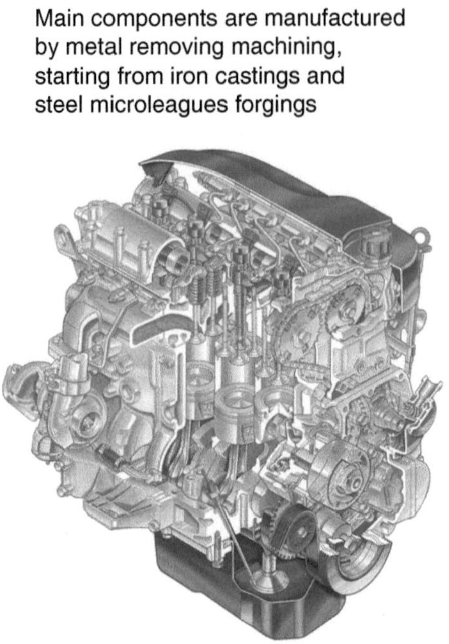

Fig. 1.15 Engine final assembling

the part of the producers. Factors that most influence the processes of final assembly are: direct manpower, material handling, specific logistic information technology systems applied (Fig. 1.15).

According to modern "make-or-buy" policies, the overall assembly and sub-assemblies are manufactured in engine assembling plants, together with engine blocks, heads and shaft machining.

For big series engines, with a modular structure, block machining is often performed on *flexible transfer lines or interconnected machine centres*. Particularly important for quality level are the boring and super-finishing operations on the engine shaft pivot location and head binding plane milling.

The machining of the head cylinder is often set-up in *machining centres* working at high speed and automatically interconnected to each other, with the ability to produce specific versions of engines belonging to the same family in a random way. Particularly important for quality level are the locations of shaft distribution and valve boring operations.

For the machining of the engine shaft, specific and precise machine tools, interconnected with each other, are used. For superficial gardening of ground and connecting rod pivots, on-line induction or carbon-nitrate treatments in heating ovens with a controlled atmosphere are used. Diameter and form ground and rod pivot precision and either dynamic regulation are very influential for the level of function quality.

Head cylinder and short block (cylinder block with movement parts) subassembly is mostly effected by automation and integrated with cold testing operations: fluid leakage test, right rotation of movement parts, right distribution of valve opening and closing phases. In this way, it is possible to avoid systematic hot testing in the power speed, which is performed only statistically.

The assembly operations of selected pistons and half-bearing coupling for engine shaft pivot location and even fuel rail and combustion rooms testing are very influential for the level of function quality, especially for diesel with direct injection "common-rail".

Final assembly of the complete engine includes assembly of auxiliary parts and relative activation controls, engine power supply and electric module integration; these operations are normally performed manually on stop and go flow lines or on parallel independent modules, supplying component assembly kits by automatic guided vehicles (AGV).

For car engines that cover a generally wide range of models, the produceable volumes/investment ratio reaches optimal conditions within an interval of 80/120 series/hour production capacity. Otherwise, for commercial and industrial vehicle engines, this ratio is included in 40/60, even in relation to their major dimensions and to consistent diversification of the product. (Also taking agricultural, nautical and construction machines into consideration.)

Depreciation significantly influences transformation cost, and for this reason it is very important to reach the maximum degree of systems utilization (Fig. 1.16).

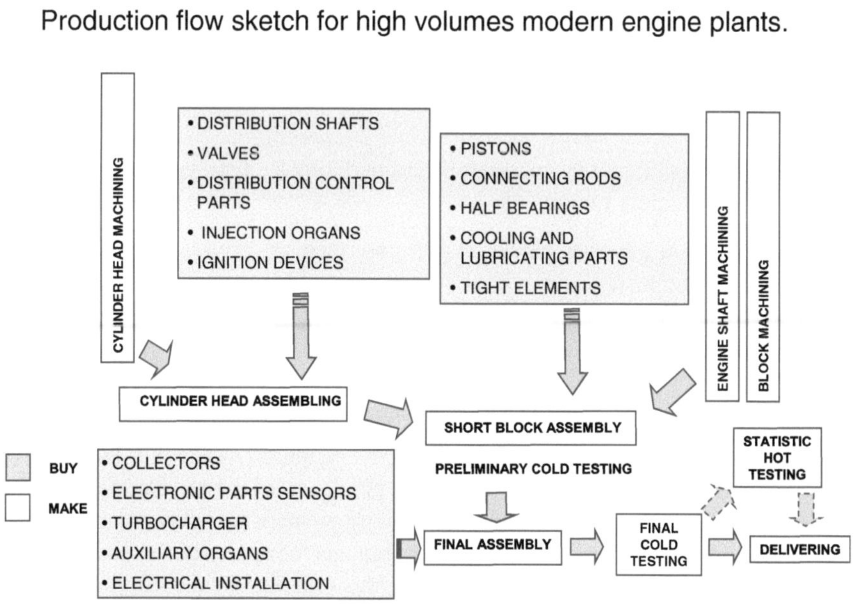

Fig. 1.16 Engine assembling process flow

The modern plant concept considers the making of parts: structural parts characterized by architecture and kinematic parts strictly related to the assembly of systems and requiring "on line" functional tests. The final producer delegates portions to specialized suppliers able to supply the total powertrain systems market, with the following parts made by hard automation production systems:

- pistons with tight rings (requiring metallurgic and finishing highly specialized technologies);
- connecting rods in micro-leagued or forged steel (requiring forming technologies with high precision);
- cam axles in spheroid cast iron or steel composition;
- special metallic league bearings;
- valves and related opening and closing control components;
- injectors/collector groups and other components for fuel supply;
- suction collectors and butterfly parts;
- exhaust collectors;
  - controlled ignition devices;
  - turbo compressors;
- flywheels and torsion bumpers;
  - ubricating system component;
  - cooling system components;
  - air conditioning compressors;
  - belts, chains and pulley;
  - composite elements for tight fluidics;
  - electrical devices;
  - temperature, pressure and oil level sensors.

Intermediate manufactured parts are farmed out to suppliers specializing in metallic league transformation:

- aluminium foundries for cylinder heads and blocks (car engines for upper intermediate product range);
  - iron cast foundries for blocks (lower intermediate product range cars and industrial engines);
  - special iron cast foundries for engine shaft (car engines of lower intermediate product range);
  - forging centres specializing in micro-leagued steel engine shafts (car engines for upper intermediate product range and industrial engines).

To manage the product's complexity effectively, and to assure constant control of the level of quality and overall equipment efficiency of the production systems, it is necessary to have information technology systems for flow management, overall equipment efficiency and quality parameter control and for assuring the traceability of those components influencing the vehicles' availability and safety level (Fig. 1.17).

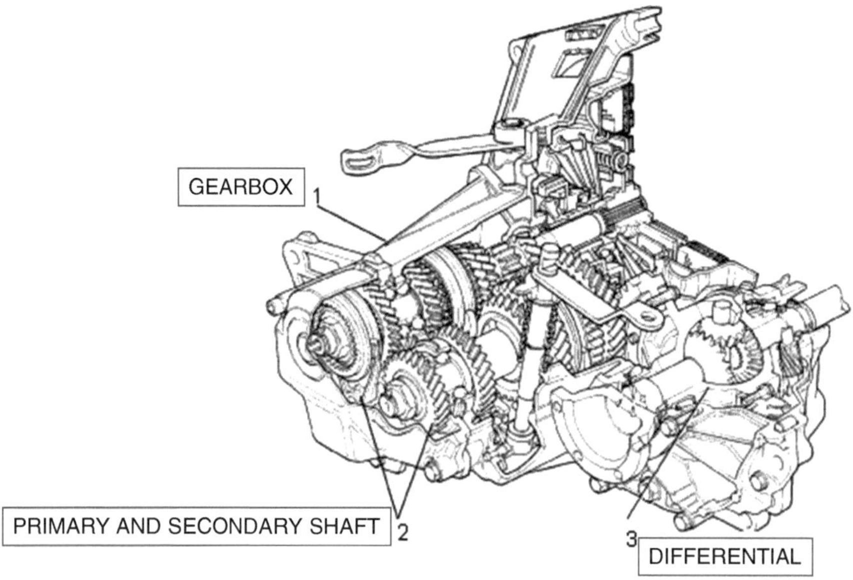

Fig. 1.17 Forward traction engine transmission final assembling

According to the organization of the modern process, engine transmission shop floors machine the following parts:

- gear shafts;
- sliding cylindrical gears;
- differential shaft;
- aluminium or iron cast parts, for shaft/gear support or engine interface.

Shaft and gear machining is organized in flow lines, with an automatic handling system to fulfil high quality levels.

Machine tools are highly specialized (numeric control lathes, precision boring and rectifying machines, gear cutting machines, finishing for shaving or profile rectifying machines…). The thermal treatment (carbon–nitrogen) is done with a reformer on pressurized equipment with a controlled atmosphere, organized on-line with the machining operations.

For product quality, it is necessary to exercise constant control over the metallurgic parameters of forged steels and extrusions (structures homogenized and easily hardening).

Technical management of forming tools is determined to obtain the necessary level of noise reduction in movement transmission (precision of evolving part profiles).

The preparation of synchro and clutch gears is normally performed in the assembly cell by manual operation, while the assembly of complete subgroups is generally automatically performed by robot and automatic screwing systems.

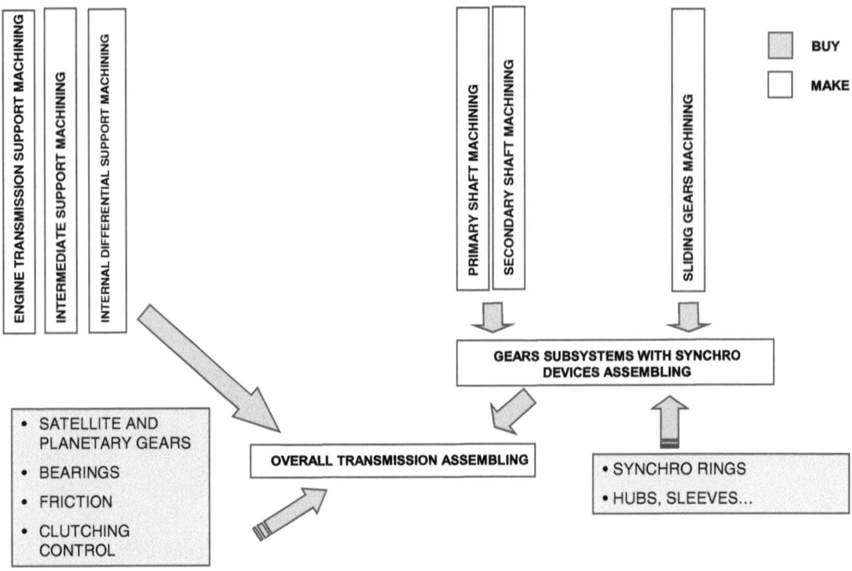

Fig. 1.18 Forward traction engine transmission final assembling process flow

Functional testing cells are equipped with sensors and automatic controls, which help to control the noise level of gears and the functionality of clutches.

The manufacturing volume/investment ratio produces its best values when standard production capacity is set at 100/140 series/hour depending on complete dimensions of the subgroups, considering that the same family of transmission could be assembled on more vehicle models, according to upper torque limit.

Machine depreciation and tool usage strongly influence transformation cost. For this reason, it is important to obtain the best degree of utilization for machines and optimal effectiveness for tools. Finally, owing to the employment of transmission on more vehicle models, production systems must be engineered to manage multiple applications (operative flexibility) (Fig. 1.18).

The final speed transmission producer delegates portions to highly specialized manufacturers, able to supply the following components to the whole transmission market with a high level of technology:

- bearings;
- differential satellites and planets (the evolving profile of which is obtained by cold printing and coining);
- speed synchronization and clutching devices;
- friction module;
- automatic controls (for automatic transmission solutions).

Intermediate machining parts are delegated to suppliers specializing in metal transformation:

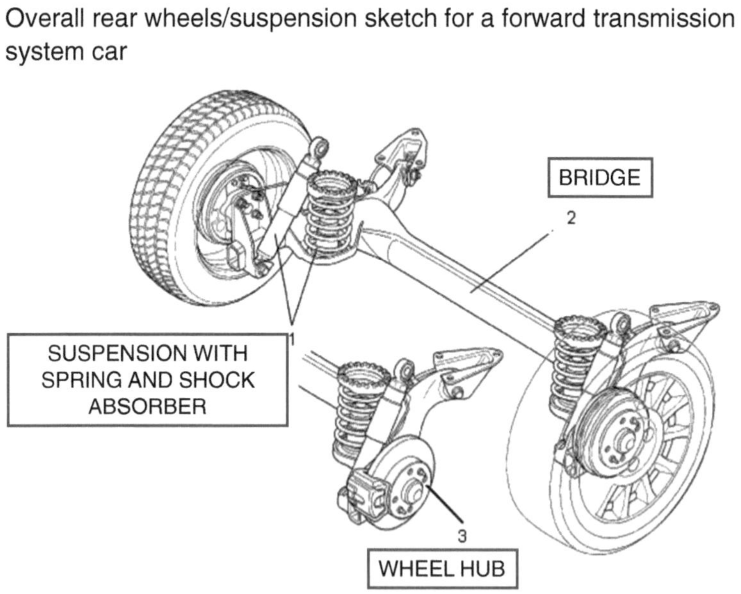

Fig. 1.19 Suspension final assembly

- precise steel forging and cold extrusion centres;
- aluminium foundries for die-casting elements;
- iron cast foundries for differential supports.

For vehicle motion safety components, it is mandatory to provide traceability for supplied batches, assuring responsibility on the part of producers (Fig. 1.19).

According to the organization of the modern process, for front and rear suspension, component machining shops are used:

- for adjustable spindles, wheel hubs and brake disks, special machine tools for high precision turning and boring, to eliminate the need for rectifying operations as much as possible;
- for connecting parts (wheels/suspension/space frame with specific solutions depending on space frame architecture) machining centres and welding systems with robot, to produce elements of equal geometry, even if they are dimensionally different (jambs, crossbars, connecting harms, twisting bridges…);
- steel, iron cast and aluminium intermediate machined parts must satisfy established safety requirements (zero structural defects).

Wheel/brakes group assembly, including jamb on adjustable spindle, is done in this machining shop by automatic systems, while the overall assembly is generally located near the vehicles' general assembly shop (decking), to minimize logistic costs and to facilitate the organization of production flow.

Highly specialized suppliers are used for:

- transmission shafts and joints
- steering wheel elements
- braking controls
- springs and shock absorbers
- sensors and software/hardware modules for vehicle braking and stability electronic control

For vehicle motion safety components, it is mandatory to provide traceability for supplied batches, assuring responsibility on the part of the producers.

The manufacturing volume/investment ratio produces its best values when standard production capacity is set at 100/140 series/hour, working on the same family of space frames, used for more vehicle models.

Depreciations are the biggest part of transformation cost and for this reason it is important to obtain the best degree of utilization for the equipment.

1.7 Systems' Strategic Prerogatives

We are going to end this chapter by pointing out the strategic prerogatives necessary for the "manufacturing systems", according to investment and depreciation plans. To compare possible alternative solutions, during the manufacturing engineering planning phase and selection of equipment and specific tools and setting for the technological layout (for simplicity, herein referred to as "systems"), it is necessary to evaluate the following "key strategic indicators":

1. *Investment productivity and utilization*
   A primary indicator is investment productivity, which is the ratio between annual available production capacity and the total amount of expenses necessary for the development, acquisition, installation and test runs of the "systems".
   A secondary indicator is the degree of production capacity utilization, which is the ratio between the total quantity of product that could potentially be produced and sold and the total amount of expenses necessary for the development, acquisition, installation and test runs of the "systems", during the normal cycle life of the investment. These analysis criteria will be examined in Chap. 3.
   In relation to *capital intensive* productions, degrees of utilization lower than 80 % would hardly allow for the obtainment of such operative results for generating value for the enterprise.
2. *Operative flexibility and investment versatility*
   Operative flexibility is determined by the producible product range (family group), relative to a specific "system". The criteria for evaluating this are developed in Chap. 3. Investment multitasking is represented by the possibility of using the investment to produce more product models/versions, and to adjust rapidly to the variability of market demand, without additional significant investment.

3. *Investment convertibility and life increase*
   There exists an attitude in a certain "system" that allows for the introduction of new products merely by substituting specific tooling and changing process automation software. Let us define the following indicators of "system" changeability:

   3.1 the ratio between the amount required for the retooling and the total initial amount spent for development of the entire "system";
   3.2 period of production stoppage needed to introduce a new model or product;
   3.3 capacity to "phase-out" and "phase-in" product simultaneously;
   For powertrain and body manufacturing "systems" in which indicator 3.1 is less than 25 %, time 3.2 is less than one month and there exists the possibility of simultaneous action such as in point 3.3, the "system" can be considered "easily convertible". By increasing indicators 3.1 and 3.2, the advantages deriving from this prerogative would obviously be decreased.

4. *Investment modularity and graduality*
   We can define "modular" as a technological lay-out made by two or more component machines or group assembly units, arranged parallel to each other and each able to produce a share of the total production required. In this situation, the "system" can be realized by steps, according to a regulator plan, with the possibility of adjusting the amount of investment in relation to the actual production level necessary while production is taking place. The biggest two factors are the modularity of the system and the total investment, considering the same automation level and production capacity required at normal production speed; conversely, this can allow for the consideration of more gradual economic obligation, reducing the "volume risk" derived from uncertainty arising from market trends and the success of products.

The prerogatives of points 2, 3 and 4 directly influence the achievement of the prerogative of point 1.

Chapter 2
From Project to Product

2.1 Standardization Logic and Project Set-Up

In industrial compartment, a "standard" is a product or process technical solution exactly defined, successfully tested and usefully repeatable by application.

The availability of a certain set of standard solutions helps the enterprise to:

(a) consolidate "best practices", increasing quality and reliability levels, reducing times and costs to develop new product and manufacturing systems;
(b) obtain economy scales in the productive process and in the purchasing of materials and working means;
(c) determine the interchangeability of parts, so as to facilitate repair operations on products and manufacturing systems and product maintenance costs;
(d) reduce work in the process of internal and external flows.

Standardization fields could be:

1. **Institutional Technical Regulation** by International Government Departments, with the participation of producer and user associations (ISO, UNI, ASA, CUNA, ETSI…). In this field: technical languages and measurements are made uniform, basic material and the characteristics of common parts are unambiguously defined, assuring interchangeability, quality certification methodologies and product and working means approval procedures are normalized, also considering user/customer and environmental policies.
2. **Product Standardization**, in the same enterprise or industrial group: design elements are defined, as to what application can be extended to a wide range of products. Furthermore, model standard solutions are defined, so that they can be easily transferred from project to project.
3. **Industrial Process Standardization**, even for the same enterprise or industrial group, to consolidate and spread out the "best practices" in their own internal processes. In this field, we have: design and testing of products and working means, control plans for materials and transformation processes, and determination and assembly of sub-assembled parts usefully transferable from project to project.

M. Gobetto, *Operations Management in Automotive Industries*,
Springer Series in Advanced Manufacturing, DOI: 10.1007/978-94-007-7593-0_2,

4. **Information Technology Systems**, set for an enterprise or an industrial group, considering homogeneous solutions in hardware and software for product design (CAD/CAE), manufacturing systems engineering (CAD/CAM/CAPE), and logistics management (PDM, MPS, MRP systems). It is also important to adopt a standard administrative procedure to simplify and speed up the accounting systems. Application of homogeneous information technology systems helps to reduce fixed costs of the enterprise, thanks also to the economic scale effect obtainable by using software and hardware solutions.

In the second and third standardization items in particular, consistent know-how is required, allowing enterprises to grow, both internally and towards partnerships and joint enterprise.

To promote the standardization process, management must focus on the consolidation of "best practices" without obstructing the introduction of new technological and organizational solutions that could offer competitive advantages for the enterprise.

Innovation must be matched by standardization. Standard solutions must not be considered "prohibitive", but "progressive", according to the logic of continuous improvement of product and process.

In the strategic setting of new models belonging to more *brands* of the same industrial group, it has to be decided which product modules can be most conveniently standardized, for reasons of economic scale, and what must be developed specifically, in regard to the distinctiveness of individual brands (style shapes, engine and space-frame solutions…).

For "mass produced" products, there is a prerogative push towards standardization, while for "premium" products, there is a prerogative push towards "distinctiveness", without renouncing standard component solutions that are not perceived by the customer as "distinguishing brand originality" and, for this reason, are usefully derived from a larger productive scale.

Let us now look at the general guidelines for setting product design for *automotive products.*

First of all, the new product's performance targets and functional and style requirements are set and are carefully compared to similar existing products, including those of the company's foremost competitors. The best-fitting materials and best technologies are chosen, starting from the standard solutions available based on company know-how. Geometric shapes and structural composition description sketches are prepared. For each module component of the final assembled product, specific parts are distinguished with a "new part number", taken from those respectively belonging to the "common component database", company-owned, or purchased from specialized suppliers chosen to be *co-designers.*

These are preliminary studies in a setting phase (pre-engineering) and they end with the institution of "technical specifications" relative to the product to be sold on market and their components. Documents and input data necessary for the development of the executive project normally include:

(1) Manufacturing characteristics of the final product, relative subassemblies and functional modules, through blocks schemes and 3D sketches representative of the product's technical architecture and macro-composition **(Product Macro-Structure = PMS)**.
(2) Performance targets and functional and aesthetic requirements, relative to the products destined for market and consequently for each of the functional modules/subsystems included in the PMS **(Product Function Deployment = PFD; Quality Requirements = QR)**.
(3) **Cost targets,** relative to the final product in each of its versions and to each item included in the PMS (unitary production costs, foreseen for the running of normal production).
(4) **Project development expenses, specific investments for industrialization, costs for production ramp-up and commercial launch for new models**.
(5) **Profit targets,** relative to the final product in each of its versions (unit prizes without considering distribution cost, sales volume within the industrial life cycle).
(6) **"Time to market" targets**.

The decision-making process normally applied for the creation of executive projects, verifying coherence with targets, matches the scheme detailed in Sect. 1.4.

We should point out that PFD performance targets are set using classic indicators for the setting up of powertrains, bodies and vehicle systems. Other target parameters derive from approved norms for safety, fuel consumption and exhaustion emission minimization.

During the setting up phase of a project, target parameters for customer quality, life cycle, maintainability and consequent warranty costs are always specified. For this purpose, *carmakers* apply evaluation methodologies based on experience and according to criteria adopted by institutionalized internationally qualified Technical Departments.

The auditing process for fulfilling the above targets is normally performed by the Quality Assurance Department, as delegated by top management, for the purpose of having an objective judgment based on comparison with competitors' products and to be sure that the root causes are correctly identified and that this information is passed on to the right company department or suppliers.

For the technical–economic part of the project, a specific activity of value/cost analysis for each item of the PMS is undertaken. The specific cost is referred to each module/subsystem. Adding the estimated cost of operation and for direct material required to complete the vehicle process (body painting, final assembly and testing), the cost target of the final product is determined. Value/cost analysis is done preventively (during the setting up phase) and continues during the executive phases in an analytical way, through the end of the industrialization process. Project Responsible has to search for necessary corrections and put them into action, collaborating with *Manufacturing* and *Purchasing* Responsible.

Even in the subsequent set-up for the executive phase of the project, it is necessary to ensure coherence with the above points 2, 3, 4 and 5. It is clear, for instance, that the sale price of the product depends on its value (i.e., on PFD and QR indicators), just as sales volume (market share) depends on the comparison of the prices to those of competitors. Investment requirement and the process of obtaining capital depend consequently on sales volume.

In modern industrial automotive organizations, these important coherence tests are performed by a specific top management staff, in charge of Product Development Planning. This department, in collaboration with Marketing and Brand and other industrial departments for powertrain and vehicle development, must design the Product Range Plan, assuring synergy and defining the appropriate level of standard carry-over and specific solutions for modules and components.

Finally, Finance and Administration has to perform an economic and financial analysis in industrial initiatives and an audit on profit results.

2.2 Industrialization Process Description

Once we have identified the project's standardization logic and actual organizational criteria applied for technical/economic checks, we have to examine methodologies used to **decrease** the ***"time-to-market"*** **of the new product, assuring contemporary PFD and QR targets**.

The block diagram of Fig. 2.1 represents all macro phases of activity necessary to develop and manufacture a new vehicle model. The upper part of the scheme describes activities for design, development and technical testing, starting from the product "concept" up to final project validation. The lower part of the scheme describes product industrialization, starting from the checks for manufacturing feasibility on the products, up to qualification of the productive process and series production start-up.

This scheme is not representative of temporary lead time of the single macro-phases, but gives evidence of the priorities between the activities and underlines possible concurrent activities, useful for reducing the "developing critical path".

New vehicle model project development and industrialization key points are:

(1) ***concept and style delivery (pre-engineering conclusion)***;
(2) ***technical validation of project (final design release)***;
(3) ***qualification of productive processes and "buy" elements (process validation);***
(4) ***delivery for production (final product validation);***
(5) ***commercial launch (delivery to market).***

The "time-to-market" corresponds with the period of time between concept and style delivery (key point #1) and commercial launch (key point #4).

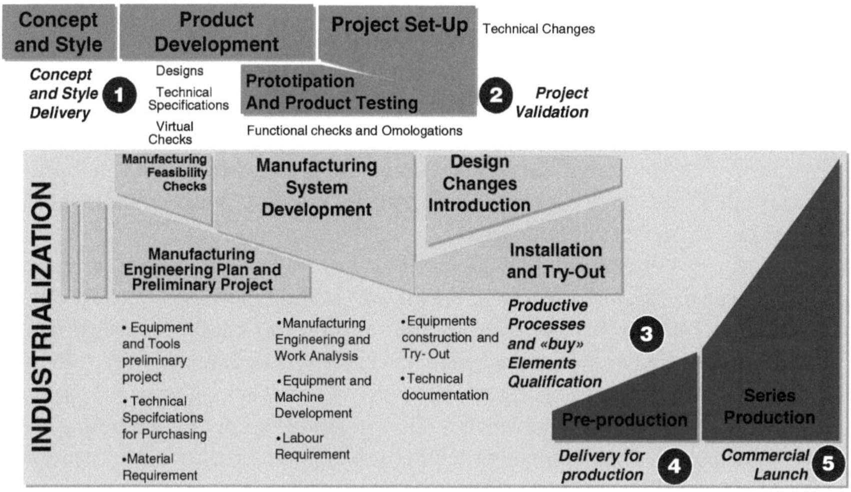

Fig. 2.1 Automotive industrialization diagram

According to the best practices of the automotive industry, to develop a new model of car, already equipped with available engines and transmissions, it is possible to reach a "time-to-market" of less than twenty-four or a maximum of twenty-eight months, depending upon the technical complexity of the product. These terms correspond to a *lead time* of eighteen to twenty-two months, depending upon the period between PFD and concept and the definition of style specifications and start-up of production (*job one*).

In the case of a new solution for the engine or space frame, the above "time-to-market" could turn out to be some months longer, due to the necessity of a highly meticulous testing phase for assuring reliability of the new product and obtaining technical approvals (as the result of more severe restrictions).

Similar consideration can also be given to the development of new models of commercial and industrial vehicles.

Let us consider modern criteria for reducing "time-to-market" and assuring in the meantime the necessary quality level from the first delivery to customers:

1. More activities are run in parallel, when possible and convenient, applying modern techniques of **grid planning.**
   To do that, **"simultaneous engineering"** is applied between Product and Process Engineering, involving partner suppliers in **"co-design"** being put in charge of manufacturing the principal components and tools.
2. During the design phase, CAD/CAE support are run, applying accurate predictive analysis, by using modern **techniques of numeric calculation for full elements (FEM) and simulation of behavior of product during usage** (shock absorption, vibration and dynamic effects, thermal effects…).

3. During the development of specific tools, modern techniques CAD/CAPE are used together with **material transformation processes and component assembly simulations**.
   The above-mentioned checks and consequent project reviews have to be done in advance in spite of the significant manufacture of prototypes to reduce critical events detectable during experimentation in the laboratory and on the road. In this way, the total number of requisite prototypes and experimentation cycles can be drastically reduced.
4. Significant prototypes are built up quickly, using material according to geometrical specifications reported in drawings; for these purposes, modern **"fast-tooling" techniques for the construction of pilot tools and representative of the manufacturing engineering plan** are used. Proceeding this way, products can be tested with accuracy, obtaining approval and reaching project validation on time **(key point #2),** in advance of the availability of definitive tooling. As a consequence, the total amount of technical design changes required decreases in the final product and process set-up phases.
5. A **pre-production phase** is run, using definitive tools, even if the try-out phase has still not been completed, so as to organize the product and process checks in a statistical data base, assuring quality levels required for commercial launch **(key point #3).**

The industrialization process ends with the "delivery for production" (key point #4), assuring the stock of final product necessary for commercial launch (key point #5). For the purpose of reducing "time-to-market", production management will attempt to accelerate production ramp-up, assuring defined quality levels.

Let us see now in detail what the necessary phases are for developing the competitiveness of industrial products.

The first phase consists of setting up a manufacturing engineering plan for *making* parts and researching partner suppliers for *buying* parts. As we will see in Chap. 3, starting from the manufacturing engineering plan, production systems are developed, choosing convenient machinery and equipment types in relation to product characteristics and the planned level of activity.

Interaction between product development and manufacturing engineering, rightly extended to first level suppliers, allows the manufacturer to:

- **verify design feasibility in advance, in relation to quality, productivity and cost targets (design for manufacturing);**
- **join elements and complete modules for production, according to the relevant logics of technological families (family groups), to realize production with the right economic scale and necessary operative flexibility (product mix model setting facility);**
- **preventively establish "process capability" levels necessary for standard operations of the manufacturing cycle influencing product quality.**

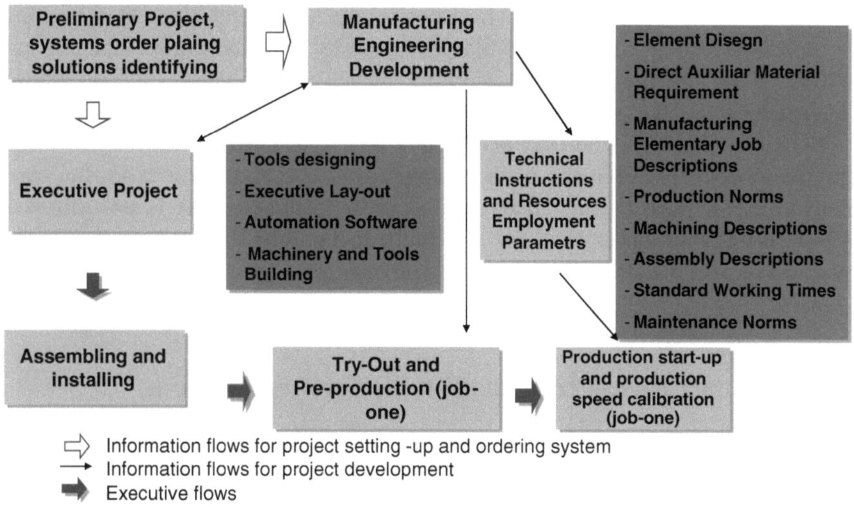

Fig. 2.2 Manufacturing system development process

The *flow chart* in Fig. 2.2 describes macro activities necessary for the development of manufacturing systems:

To reach quality and productivity targets, with the best "lead time", it is particularly useful to apply modern "Project Management" methodologies, which can be broken down as follows:

(a) activity matrix description building, so called WBS (Work Breakdown Structure);
(b) attribution of the single items of related tasks reported in WBS, underlining necessary co-operations for "simultaneous engineering";
(c) definition of time needed for activities, using GANTT diagrams and underlining sequence obligations and possibilities for running phases in parallel;
(d) definition of the critical path for determining project development and specific tools, both for the making and buying of parts, and total lead time;
(e) professional resources necessary in Technical Centers for each planning phase of the project.

In advance of the run of the process, before series production begins, the Product and Process Engineering Departments must provide the following items to the Manufacturing Team:

- updated technical documentation for product and process;
- technical instructions for conduction and maintenance of the machinery and for tools assigned (normally provided by equipment constructors);
- tools and parts normally subject to usage;
- specific spare parts, necessary for "back-up", in case of technical failures.

In advance of start-up of production, production and maintenance workers must be trained according to a specific plan designed by manufacturing managers and performed in collaboration with working means and suppliers of information systems.

Normally, it is allowed that productive flow speed will be initially low and only progressively reach the project target after a certain time from the start of production (due to the "learning curve"). Conversely, it is not allowed, either initially or otherwise, for products to be sold that do not fulfill the qualitative and functional customer requirements.

2.3 Product/Process Information Technology System

By the abbreviation Product Data Management (PDM), we mean the product information technology system that allows for specifying, correlating, storing and transmitting—in a digital format and by WEB applications—parts design, technical data and all information necessary for developing industrial products, in phases from engineering to after sales.

A good PDM system uses modern information technology support. Operative system and communication procedures applied must always conform within the same enterprise or enterprise group. In this way, interaction between different departments and suppliers, co-operating on the same project development, production management or commercial aspects, will be facilitated, even if physical distance is consistent.

An Integrated Information Technology System feeds the fundamental processes represented in Fig. 2.3.

PDM is the base of a modern integrated information technology system, able to connect product engineering, manufacturing engineering and production

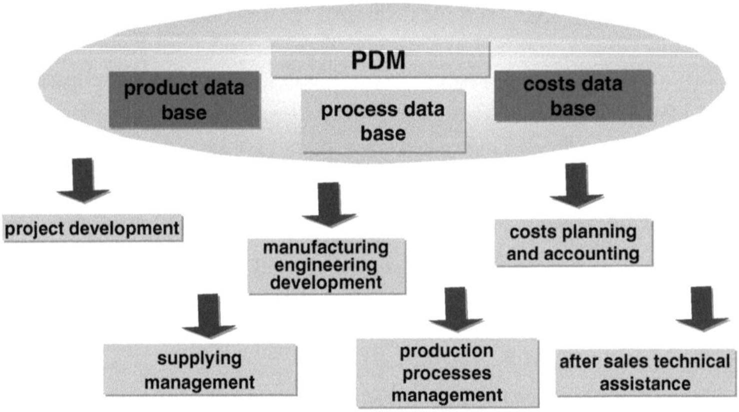

Fig. 2.3 Product data management

management activities during the product's life cycle, defining a concept of "net enterprise" (including first level suppliers, plant dealers and after sales technical centers).

Furthermore, PDM represents the informative backbone for the way in which accounting is done for single product direct costs, as we will see subsequently.

In PDM, product ***data management*** requires the operative features represented in Fig. 2.4.

Data **storage** runs as follows:

- data are structured according to a product composition tree (bill of materials) and process logistics (manufacturing job descriptions);
- all documents in electronic format are referred to reference models;
- original data (technical specifications, CAD part design…) are stored as virtual data.

Examples of virtual data:

- file typology, its position in a file system, creation data, author…;
- link between a CAD file and a text file describing it;
- raw material technical characteristics, requirement and unitary cost of each part number.

Data **research** runs as follows:

- virtual data can be used as an information source or as searching keys in the system database;
- data can be accessed through WBS visualizations.

Starting data **management and updating,** i.e. **changes**, runs as follows:

- PDM users work on original data copies (consultancy, changes);
- all changed data are recorded as revisions of the original data;
- original data always remain in memory.

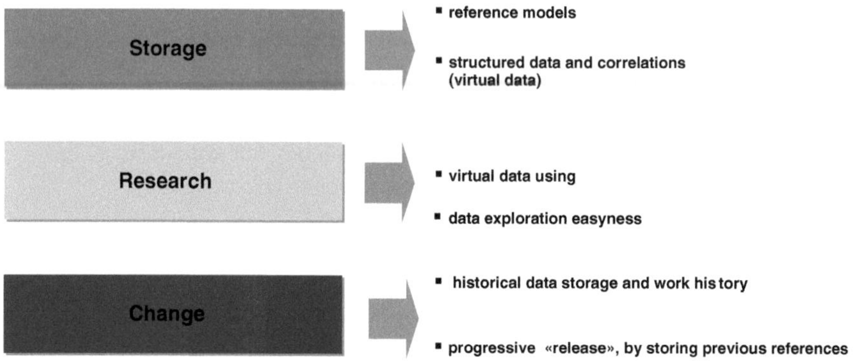

Fig. 2.4 PDM features

PDM allows for managing ***technical data and the creation of part number designs and the updating process*** of the product, checking the following operative aspects:

Work management, which means establishing *who can do what*:

- system users are classified by role, according to their position in the organization;
- specific possibilities for accessing and processing data for each role.

Workflow management, which means establishing *who must do what and when*:

- the system considers a "standard workflow" (execution, check and approval), for all the items used in the activity;
- the system automatically sends the updated data to the responsible parties, according to the above-mentioned "standard workflow" and established procedures for delivering the design's "release".

Work history management, which means establishing *who has done what and when*:

- the system tracks access to data and completed operations (release, changes, approval…);
- at every moment, it is possible to know the work history of all data (model and document).

Simultaneous engineering, which means establishing *who does what with whom*:

- the system foresees potential for collaboration between individual positions and departments;
- in this way, revisions are automatically posted (new releases), keeping the participants in the project constantly informed.

Modern PDM systems are integrated with other enterprise information technology systems, on the condition that they have homogeneous environments (same CAD/CAE/CAPE systems), as shown in Fig. 2.5.

An integrated information technology system allows PDM to deal with the other systems used in the product/process development, for:

- creating input data(design engineering settings);
- visualizing designed parts and their position on complete modules;
- updating and spreading out of data, allowing electronic research.

The main advantages of a homogeneous information technology environment are:

- access to same data from different hardware platforms;
- same CAD models and structural data available in the same environment;
- availability of most up-to-date data.

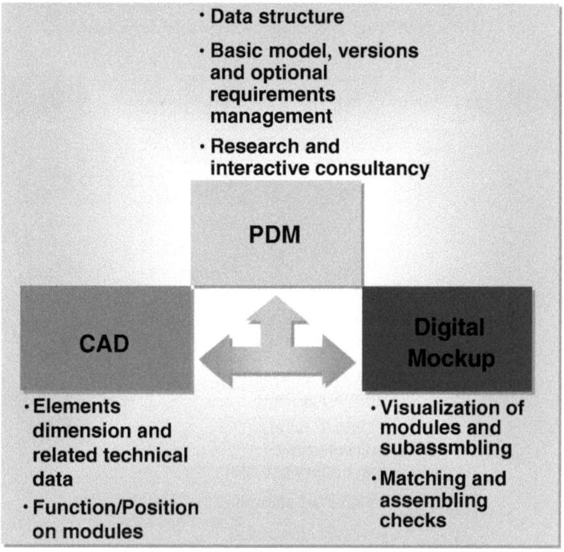

Fig. 2.5 PDM integrated information technology system

Internet and Intranet connections allow for the elimination of geographical restrictions and improve interaction, with the application of necessary protections for technical documents, according to policies established between provider and supplier.

The heart of PDM is the **Product Breakdown Structure** (**PBS**), described in the following section.

PBS structure is *hierarchical* and is represented in Fig. 2.6 (tree diagram). The final product is on the top and is deployed, through different levels, down to single components, subassemblies and raw materials at the base.

Automatic data processing, according to the tree diagram, means assigning a code to each link on the structure; through the code, each part is associated with the assemblies and subassemblies to which it belongs.

In describing ***PBS*** function, we will use the following terms:

- ***element*** = design code or part number;
- ***assembly or subassembly*** = secondary group code
- ***usage coefficient*** = number of elements necessary to complete a final product
- ***completed elements and final group weight*** = it corresponds to element designs and is calculated based on volume and weight of direct materials, including the fluid ones (metal external parts and polymers, paints, lubricating, structural glues and sealers…).

This data, referred to as the several typologies of basic materials, is necessary for statistical control and technical and economic analysis: comparisons between theoretical and real data, comparisons between models, benchmarking between different solutions, cost effect analysis as a consequence of raw material price variations…

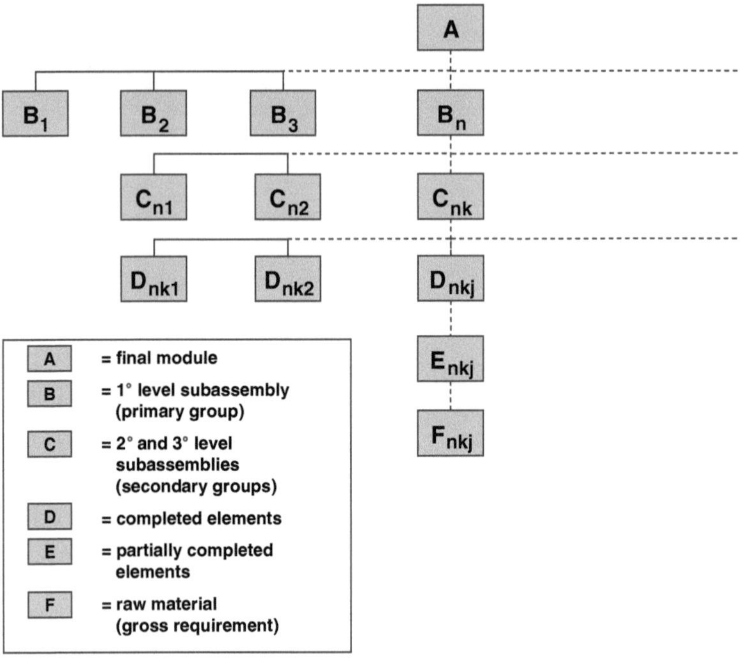

Fig. 2.6 PBS Hierarchical structure

Some significant examples:

- *powertrain system*, the completion of body brackets is a primary assembly of level B, but when installed on the vehicle, it is considered to be of level A.
- at level C, we find the principle mechanics groups included in the powertrain system: *complete engine, transmission, gearbox, electrical systems, brackets for connections to the body.*
- remaining on the complete engine, at level D, we find the engine short block (including *cylinders, blocks, engine shaft, con rods and pistons…*) and the head cylinder group, including valves.
- then, at level E, we find different primary groups and subgroups: *head cylinder, pistons, valves, injectors, inlet and exhaust manifolds, flywheel, auxiliary controls and fixing elements…*
- finally, at level F, we find *sub-components,* specific *half transformed elements* and raw *materials* for each of the above elements.

Even as an example, *material requirement* for an exhaust valve is the part of a steel bar used for hot printing and the quantity of coating reported on the cone valve before finishing the operation. If, for example, the engine is a four-cylinder, with four valves for each combustion room, the *usage coefficient* for the element exhaustion valve is eight for each product series.

Despite being unambiguous, the structure of a car model is developed by steps: "macrostructure" is determined during the project setting phase, "analytical structure" during the technical project development phase and "bill of materials" during the manufacturing engineering phase. This classification is typical of automotive processes, depending on the complexity of the vehicles.

Figure 2.7, according to the "best practices", represents the time sequence flow occurring from project set-up and industrialization to the definition and codification of the product structure and production competency.

Starting from PBS, the Project Manager, as indicated in Sect. 2.2, defines the Work Breakdown Structure (WBS) and controls project development, assuring that:

- professional resources and technical structures will be available on time;
- value/cost targets will be reached, according to PMS items;
- defined times and motions for each activity and phase will be respected, in relation to time to market, with special reference to the "key process points" mentioned in this section.

PMS being organized in "principal modules and functional subassemblies", it is not necessarily responding to physical aggregation of the product.

Some examples of "functional subassemblies" as single items of **PMS** that are related to several "physical modules" of the product, according to assembly procedures:

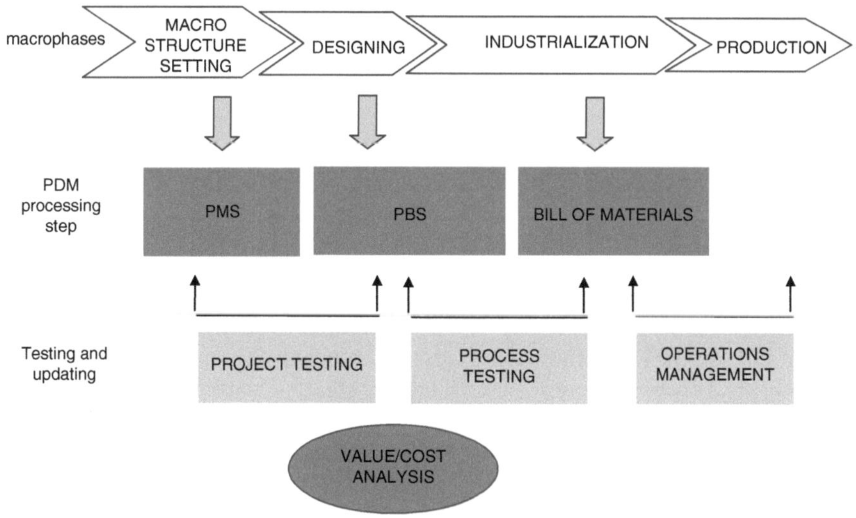

Fig. 2.7 Product structure codification process

- engine supply system, partially allocated to the engine and partially installed in the fuel tank and on the vehicle;
- engine cooling system, partially allocated to the engine and partially installed on the "front-end" of the vehicle;
- steering wheel system, partially allocated to the front suspension module and partially directly installed on the vehicle;
- brakes controlling system, partially allocated to the wheels group (corner) and partially directly installed on the vehicle;
- air conditioning system, partially allocated to the engine (compressor) and partially installed in cockpit module and directly on the vehicle;
- electrical equipment (wirings and electronic switchboards);
- electronic devices for engine, brakes and vehicle suspension control…

It is important to note that for each element design in the **PBS** a relative function code can be assigned, according to the **PMS**. As the project is developed, each "principal module and functional subassembly" becomes an "element design", which is described in the product breakdown structure, applying classical criteria of "equipment construction".

Practically, **PBS** accurately describes the product's physical structure and is extended to 1st level suppliers and their sub-suppliers, according to "make-or-buy" competency.

Using the **PBS**, it is possible to:

(a) manage the designs and standard carry-over solutions with other product lines;
(b) underline links between elements/subassemblies and final product;
(c) establish precise "make-or-buy" competency for each element;
(d) activate manufacturing engineering plan for the "making" of parts;
(e) activate purchasing orders for the "buying" of parts;
(f) confirm costs for module components, considering the targets defined in the setting phase, according to the **PMS**.

PBS relative to product range items are the basic information of the PDM system mentioned above.

Once the manufacturing engineering plan for "making" parts and purchasing orders for "buying" parts have been completed, it is possible to compare analytical data derived from the executive project, with the global cost target of the product, established in the economic initiative of industrialization. Gap analysis is managed according to the **PMS**, for allocating responsibilities to the single project leaders and *co-design* suppliers, consequently leading the countermeasure research on running. Normally, the Project Manager has a safety margin for changes and unexpected events, outside of PMS, for what he could do in consequence of documented needs.

The **Bill Of Material (BOM)** follows PBS (and, in many cases, is the same document) and admits elements and subassemblies (physical modules) in consideration of *making* processes and manufacturing procedures for subassemblies and final modules. The information in it is related to a specific product (series of elements, subassemblies and assemblies) charged to a single plant and includes:

- "design code" relative to a single product element, subassembly and final assembly managed by the shop floor;
- eventual alternative design codes for the same function;
- use of coefficient of the single element, for the same product series;
- "making" or "buying" competency, shop in charge of manufacturing, department in charge of supplying and direct material receiving;
- quantitative and qualitative specifications for raw material and components necessary for the "making" process (identification code of the material, gross requirement for the single element and use coefficient during transformation process).

Moreover, along with the data and correlations specified in the PBS, the bill of material also includes the ***raw material and pre-machined parts*** necessary for manufacturing each element, including waste of material consequent to the transformation process and not directly recoverable in the same process (gross unitary requirement).

This data refers to the several typologies of materials as so-called "undefined materials", or half machined parts with a specific design. For undefined materials, the **standard requirement** is expressed in weight unit, or geometrical unit, depending on the features of the transformation processes. For example, in foundries, the weight unit is used for the single "shoots", without waste recovered in the same process, in metal printing, surface unit referring to a specific thickness of the sheet of metal is used (convertible in a weight unit without the waste of metal re-used to print other parts), and for extrusions and laminated parts, a length unit referring to a specific section is used (also convertible in a weight unit).

For all these reasons, the bill of material, despite its name, is more than a list, having a wide variety of information.

The standard requirement, as mentioned above, related to the purchasing price and to the utilization coefficient, allows for the allocation of the direct material costs to the single element produced and thus the cost of each finished element to the final product, applying the utilization coefficients to the elements defined in the bill of material.

The difference between the real consumption of raw material and the standard requirement is due to the losses caused by process failures and production scraps.

In Fig. 2.8, the logical productive flow for the *BOM* is represented.

The *bill of materials* has to be updated constantly during the entire life of the industrial product cycle, as it allows us to:

(a) manage production planning and final product delivery to the customers;
(b) manage the "completely knocked down" (so called CKD) elements for the Assembling Plant operating in "co-makership";

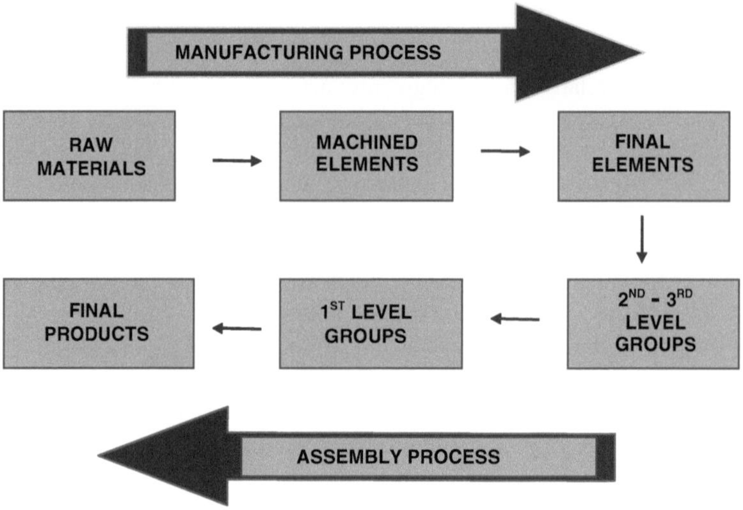

Fig. 2.8 BOM productive flow

(c) manage the process of order emission to the suppliers and the direct material requirement process (MRP procedures);
(d) manage spare parts and after sales assistance;
(e) check the utilization degree of direct materials;
(f) identify direct material in input from the final product in output, feeding the accountability financial system, for the billing as well as for stock management;
(g) assign the "cost centres", the standard working times for each element, sub-assembly and assembly, complying with the standard operative procedures;
(h) aggregate the production costs for each element and consequently assign them to the final product.

Through the PDM system, which normally works on a "PC-based" system with a network connection, it is easy to access all the technical documentation of the project (element designs, assembly schemes, material technical specifications....) and elaborate documents necessary for product management. It is possible as well to elaborate the manufacturing engineering documentation (standard operative procedures, control instructions...) necessary for process management.

Let us end this section by showing the chart in Fig. 2.9 as to how the database can be connected on a wide band system throughout the company intranet.

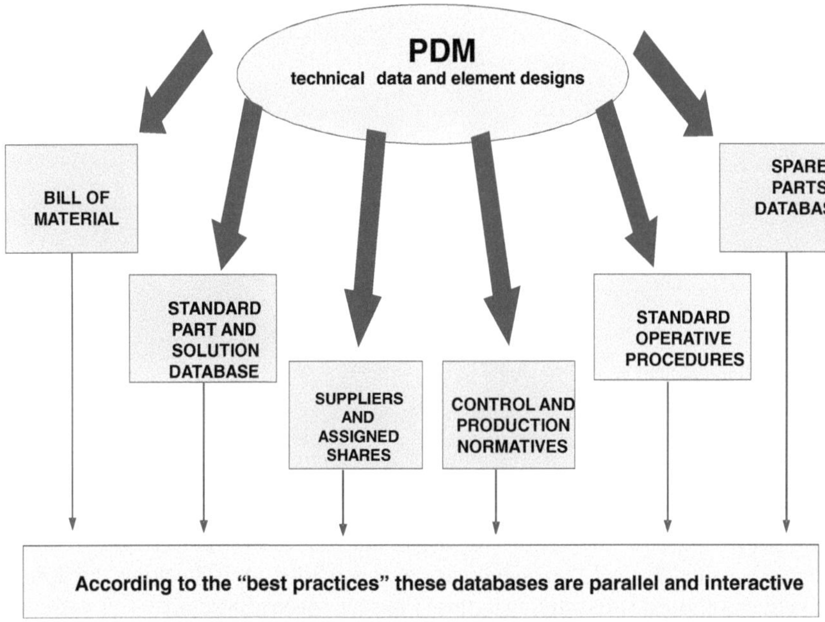

Fig. 2.9 Product process integrated system

2.4 Product Composition Analysis

To give an idea of the complexity of the projects and productive processes, in relation to the number of parts (element design) and the number of levels of composition (sub-assemblies, final assemblies), we keep in mind what follows:

- the PMS for a car model typically includes a number of 150/300 items, depending on the category and complexity of the product;
- the *PBS* is made by 2000/4000 items for the final carmaker, while it becomes at least ten times bigger if we also consider mechanical groups and the number of components managed by the suppliers along the supply chain;
- the *BOM,* on which the material requirement planning and the production "scheduling" is based, is made on a deployment concept and focuses on a single plant and production shop included in the supply chain, at the single phases of the integrated production process.

The methodological aspects related to the supplying of direct material and the final dispatching and delivering of products will be detailed in Chap. 6, dedicated to the logistic processes.

Without the actual information technology systems, it will be very hard to manage the amount and the complexity of the element designs applied to the automotive process.

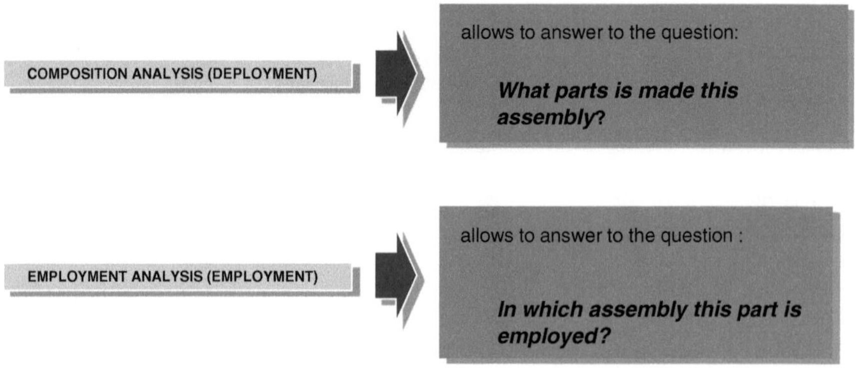

Fig. 2.10 PDM analysis

We now show the ***deployment and employment analysis,*** made possible by consulting the PDM records (Fig. 2.10):

The above composition analysis is important, first of all, for verifying that the designs are complete and their aggregation as a whole, also checking the coherence between the PBS and the analytical Product Macro Structure.

The other analytical purposes are:

Deployment at a single level to:

- identify the element required for each single level
- run the purchasing or warehouse withdrawal orders

Deployment for scaling down to:

- elaborate the manufacturing standard procedures or maintenance manuals
- elaborate the spare parts manuals

Deployment for resuming to:

- establish the components requirement based on the employment coefficient and the production plan
- analyze the component, sub-assembly and final assembly costs

Deployment of raw materials to:

- assign the basic material and component replenishment programs, according to the supplier shares
- plan the technical changes

Employment for scaling down to:

- analyze the employment of parts referring to subassembly and final assembly
- describe the technical changes: what subassemblies and assemblies are involved? What change of components must be related?

Employment for resuming to:

- establish the total basic materials required for the transformation process, considering the gross and net amount
- evaluate the utilization coefficient for the several typologies of basic materials analyze the influence of the variation of the cost of the parts on final product evaluate the consequence of a potential specific missing component of the production plan (model, version, optional...) so as to take the necessary countermeasures in production management.

2.5 Management of Technical Changes

In the automotive process, it is important to manage ***the technical changes of a product*** effectively.

In fact, product development and design activities require many changes during industrialization. Once designs and assembly schemes are defined, the prototyping phase is run to verify the project's validity; in this phase, there are many technical changes that become necessary for correcting the project and, as a consequence, many project revisions are required: these must be preventively evaluated so as to consider the cost/benefits and the impact on the development lead time.

This analysis becomes much more important when we are still in the development phase and the drawings have been used to design tools and equipment. In this case, it is also necessary to modify the specific tools already built.

During the development of new products, the Project Manager is responsible for assessing the true necessity of the technical changes and verifying their compatibility with the economic targets of the initiative (R&D expenses, investments, product cost) and with the "time to market" targets. Nevertheless, the decision-making process should be as quick as possible to guarantee extra additional costs and critical conditions for the "time-to-market".

For existing products, the main reasons for technical changes are as follows:

1. **product content improvement, the** consequence of specific requests from "marketing";
2. **quality level improvement for market alignment** (style, functionality, reliability), considering the same product's contents;
3. **direct material cost reduction** (raw material and "buying" components);
4. **transformation cost reduction** ("making" parts);
5. **"production non-quality" cost reduction** (reduction of rejections, reworks...);
6. **"after sales non-quality" cost reduction** (warranties).

As a consequence of actions related to point 1, we should have a profit increase (increasing of price or sales).

Actions related to points 3, 4, 5 and 6 must be preventively evaluated from an economic point of view so as to guarantee a good cost/benefit return.

Technical changes for solving non-quality problems perceived by customers have to be introduced at all stages, searching for solutions with minimum development cost and less impact on final product cost. Improvements that reduce warranty costs (ideally more than three years for the main carmakers) when increasing product cost or requiring nonspecific investments must be economically evaluated to check the cost/benefit return as well.

To finish this chapter, let us describe the criteria used to introduce the technical changes for a new product during the industrialization phase or the production phase. For this purpose, it is important to distinguish the following ***levels of complexity,*** progressing from the lowest level to the highest:

1ST LEVEL for a single element or more, without interchangeability constraints; the design number changes, but not the function code of the element, and it is sufficient to act on the specific tools of the single components.

2ND LEVEL for more elements, with congruence and interchangeability constraints; composition for assembly does not change, nor for the sub-assemblies; design number changes, but not the function code of the element, and it is sufficient to act on the specific tools of the single components.

3RD LEVEL in addition to what was referred to in the previous point, the structure of the assembly and sub-assemblies also changes; in this case, the design number changes as well as the function code of the elements. It is also necessary to review the assembly schemes and the manufacturing engineering plan.

Considering the above three levels of complexity and introducing an emergency level, it is possible to introduce a technical change in different ways, as described by the scheme in Table 2.1.

In planning the introduction of the changes from level 2 and 3, in case of modifications in tooling that prevent the production of components before the changes, it is necessary to evaluate the specific spare parts stock level for after sales assistance.

Considering what is mentioned above, the necessity of planning the introduction of technical changes, and complying with the requirements of the Product and Process Engineering Department, Manufacturing Department and Supply Chain Department, is clear (considering also, when necessary, the Purchasing Department, the After Sales Department and the Suppliers).

Once these preventive checks are completed, all instructions are invoiced to all departments involved simultaneously by e-mail. The Product Definition Department is in charge of this task; this department is also in charge of coordinating the other secondary plants for the introduction of changes in other co-producer plants.

If the "scheduling" is inconsistent, the introduction of technical changes can cause delays in the production process and/or economic losses, which can also be due to the rejection of direct material no longer usable.

Table 2.1 Technical changes management by priorities

Level of complexity and of emergency	Introduction methodology	From
1st level/not urgent	It is planned by the end of the stocks of the components before change with significant value	Progressive number detected after the introduction
1st level/ urgent	It is planned by date, speeding up the specific tooling as much as possible	Progressive number detected after the introduction
2nd level/not urgent	It is planned from a certain progressive number, after having checked the availability of the components of the same type and after the end of the stocks of the components before change with significant value	Progressive number detected after the introduction
2nd level/ urgent	It is planned by a specific date, after having checked the availability of the components of the same type and speeding up the specific tooling as much as possible	Pre-established progressive number
3rd level/not urgent	It is planned from a certain progressive number, after having checked the availability of the component of the same type and after the end of the stocks of the components before change with significant value. It is also important to check the availability of the new assembly and after sales instructions	Progressive number detected after the introduction
3rd level/ urgent	It is planned from a certain progressive number, after having checked the availability of the component of the same type and speeding up the specific tooling as much as possible. It is also important to check the availability of the new assembly and after sales instructions	Pre-established progressive number

In the automotive sector, technical changes require the setting of a significant amount of expenses; this amount influences the economic balance, and for this reason it is checked periodically by the Finance Department. The forecast of the total amount of projected expenses is foreseen in the budget and checked finally for each version of the product, deploying the differences for each one of the cost centres and for each of the root causes that have generated the changes.

With this last section on technical changes, this important chapter, dealing with the industrialization of a new product and product management during its life cycle in production, ends.

Chapter 3
Manufacturing Engineering and Equipment Efficiency Evaluation

3.1 Manufacturing Engineering Planning and Executive Project

A study of Manufacturing Engineering starts with the product and the elements to be manufactured: element designs and geometrical precisions needed, machining and basic material characteristics, breakdown of work structures for the complete subassembly, etc.

By *working cycle*, we mean the right sequence of operations needed to transform a certain basic material into a partial or full machine, applying specific transformational technologies. By *assembling cycle*, we mean the right sequence of operations for compiling components into subgroups or final product subassemblies, applying specific technologies for assembling and joining.

Criteria for tooling and automation level decisions were discussed in Sect. 1.4 of the previous chapter. It is to be emphasized that the operative speed of a process depends on its technological characteristics, equipment capacity and the physiological limits of manual activity.

3.1.1 Manufacturing Engineering Plan Study and "System" Preliminary Project

Manufacturing Systems for series production are developed starting from a preliminary manufacturing engineering study (working solutions), which allows for defining the following process targets:

(1) Operations needed, repetitiveness during the cycle, sequences and total duration (significant data);
(2) Qualitative and quantitative product output for different phases of the process, in relation to design elements and operation allocation;
(3) Typology and technical specifications of equipment and working means;

M. Gobetto, *Operations Management in Automotive Industries*, Springer Series in Advanced Manufacturing, DOI: 10.1007/978-94-007-7593-0_3,

(4) Criteria for project setting of specific tools, in relation to functional needs and chain of tolerances;
(5) Automation level and equipment typology for inter-operational connections.

For quick and efficient implementation of the above described ***manufacturing engineering plan***, it is necessary to prepare a database including basic data and standard solutions. These documentations are derived from field applications, benchmarking analysis or research projects.

The manufacturing engineering plan then leads to the preliminary project and the standard solutions delivery (technical specifications book) necessary to purchase equipment, machine tools and tooling. Technical documentation for the preliminary project includes:

- Brief manufacturing cycle, in which data derived from the manufacturing engineering plan are resumed;
- Preliminary technological layout, which features signed operative unit positions, workplaces and material flows;
- Manpower required for a functioning process;
- Plant spaces and power supply requirements, evaluation based on specific parameters;
- Investment required, evaluation based on specific parameters.

This information is fundamental for the planning of economic initiatives, elaborate technical data and the specifications book for purchasing equipment and tooling.

3.1.2 Executive Project

Starting from the *manufacturing engineering plan* and *preliminary project,* manufacturing cycle operations are analysed, and specific tools and automation systems are developed as follows:

(a) After tenders and purchasing negotiations, equipment and tool manufacturers are put in charge by the carmakers, so that it is possible to proceed simultaneously with the engineering in the subsequent project phases.
(b) Material technical specifications and geometrical shapes of input elements are deployed for each process phase. Consequently, the degree of utilization of direct materials is evaluated, considering the process rejects and over-metal necessary for the transformation process, in hopes of finding more convenient solutions. Some of the more significant examples:

  - to minimize metal rejects in shearing and shape forming processes, the best sheet metal shapes and positions are to be determined, assuring material border portions strictly necessary to keep sheets during forming operations;

- in the painting process, electrostatic systems driven by programmable robots are used, with optimized routes to assure uniformity of layers and minimize *over spray;*
- to produce differential secondary gears, an evolving conical profile is made by a precise printing process, so that over-metal may be reduced at the same time as the elements of the finishing operations are drastically reduced.

(c) Diagrams of time-sequences for each operation are analytically defined considering obtainable productivity in each process phase, as described in Sect. 3.2 below.

(d) Labour standard working time required for the entire working cycle is evaluated as described in Sects. 3.3 and 3.4.

(e) Electrical power supply and the auxiliary material consumption necessary for execution of the working cycle are preventively defined, applying criteria described in Chap. 5.

(f) Project ends with tools CAD design and elaboration of *executive layout* and *job operation descriptions.*

Based on the above, project documentations of specific tooling and equipment are defined and installed, and the manufacturing process is developed. The executive layout represents:

- Positioning of machines, equipment and devices;
- Material flow and positioning of manufacturing process input components and output products;
- Position and classification of manpower involved in working cycle;
- Buildings and installations necessary for equipment and the connection of machines to energetic and fluidic nets, including exhaustion of process materials (relative documentations are normally attached to the executive layout so that equipment delivery can be worked out in advance).

Virtual Factory techniques are used to optimize working cycles and tool the executive project: operative units and specific tools are represented in 3D, related to CAD design elements; kinematic functioning simulations for automatic devices are run to prepare for possible temporal or physical interferences during the work cycle, and an ergonomics standard is used to determine the best manpower position. Following these checks, necessary changes to the project are made to reduce time during the technical system set-up and regulation phase. For managing an executive project, it is very useful to apply the methodological criteria described in Sect. 2.1 of the previous chapter, in relation to systems WBS. As the executive project runs, it is possible to calculate analytically the unitary cost for each component, subassembly and final assembly of product, applying the methodological criteria featured in Sect. 4.5 of the following chapter. Furthermore, in advance of the purchase of equipment and tooling, based on WBS, it should be checked that necessary capital is consistent with the investment plan laid out in the project initiative.

3.2 Production Capacity Setting

In most industrial situations, the interaction between manpower and operative units and automatic systems is particularly relevant. Starting from the manufacturing engineering plan and the machinery and equipment executive layout, detailed in the following section, an analytical time description of the operations is drawn, represented by a GANTT diagram, in which micro-phases included in the operative cycle are reported. This diagram of repetitive operations is called an operation time-sequence diagram. Reported within it are:

(a) time sequences of each micro-phase, considering priorities and contemporary events;
(b) length of each micro-phase, according to optimal speed of machinery and manpower;
(c) "critical path", which determines operative cycle length Working Cycle Time (WCT).

In Fig. 3.1, a simple example of a time sequence diagram, where a semi-automatic operation for door-welding is represented with man/robot interaction, shows how a couple of phases in shadow do not contribute to the critical path that determines the Working Cycle Time.

The speed and stability of the phases of the above-mentioned "critical path" hardly influence the technical system's productivity. For each one of the phases, the following optimal conditions are theoretically and practically striven for:

- operative units' maximum working speed, according to maximum power of equipment parts, to the stress made on machined elements and on specific tools, also considering the tools' economic duration;

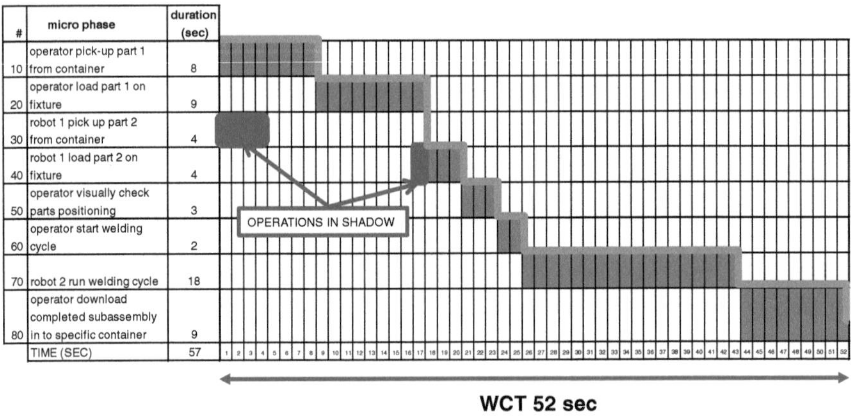

Fig. 3.1 Time sequence diagram example for a semi-automatic welding operation of a door

- minimum time required for transforming operations, according to product specifications from a quality point of view and also considering necessary "process capability" levels (this argument will be addressed in Chap. 8);
- sustainable working speed for manpower, to guarantee consistency of manual operations included in working cycle, according to criteria laid out in Sect. 3.3 below.

In the same operation time-sequence diagram, recurrent manual phases include:

- loading of direct material and unloading of final products, not performed by automatic handling partners;
- the handling of "partner" tools and the eventual manual control of operative machines, due to safety reasons;
- Assembling, finishing and controlling operations not performed by automatic systems.

Let's call *Working Cycle Time* (WCT) the result of the operation time-sequence diagram, including both manual and automatic phases. In automotive production, WCT is typically measured in minutes. The frequency of the recurrence of operations is called "pace" and it is typically expressed by the number of cycles per hour, with the exception of very high rhythm machining operations, where it is expressed by number of cycles per minute (sheet metal printing, metal forging and metallic component).

With the production process set, WCT is a standard input data. It is preventively set during the development phase and then tested and improved during the "systems" try-out phase, with the aim of reaching the project's productive capacity.

WCT determines productive flow speed, represented by the "Hourly Virtual Production": HVP = 60•k/WCT, where k is the number of elements produced for each working cycle.

In addition to the recurrent phases included in the "operation time-sequenced diagram", intervention with machines and tools can be expected with periodic frequency. Those activities are related to the following *process auxiliary operations*:

- tools with rapid usage calibration and change;
- batch production changeover;
- direct material supply container substitution;
- statistical process control for parameters affecting "process capability" and consequent set-up;
- technical recurrent cleaning;
- daily and weekly start-up and shut-down operations.

For each of the above periodic auxiliary operations, necessary for optimal production, times and methods of action are carefully analysed, searching for optimal frequencies to assure "process capability" and to obtain the best performance from the equipment.

Let us define *process unavailability index pu* as the incidence of interruptions due to the above-mentioned operations on working time, considering their length and frequency:

$\mathbf{pu} = \sum_{i=1}^{N} \left(\frac{SUUT}{P}\right)_i / \mathbf{WCT}$, where:

(SUUT) is the unavailability of the standard time for production, due to the single set-up periodical operation (i);
(P) is the required period (number of working cycles required);
(N) is the total number of periodical operations set up to perform during a working shift.

Let us now introduce "standard productive capacity" (SPC), related to a single "operative unit" or to a "multi-operational system": it defines the quantity of producible elements in a defined available working time (AWT), assuming full continuity and stability, without any technical or managerial failures, but considering the time necessary to run periodic auxiliary operations.

$$\mathbf{SPC} = \frac{60 \cdot k \cdot AWT}{WCT \cdot (1 + pu)} = \mathbf{standardproductivecapacityduringavailableworkingtime}\ (\mathrm{AWT}),$$

where:
k is the number of elements produced in each working cycle;
WCT is the working cycle time measured in m';
pu is the process unavailability index;
AWT is the available working time, measured in hours/day or hours/week.

Considering the technical reliability coefficient R of the "system", it is possible to calculate the *Available Productive Capacity*: **APC = R** · **SPC**.

WCT and *pu* are typical of the technological process; *k* is typical of tooling (strictly related to batches dimension); *R* is typical of machines/equipment and technological layouts, but also depends on maintenance management effectiveness.

Reliability coefficient *R* is determined by statistical data collecting, applying predictive analysis techniques (process FMEA) during the system development phase; during production, normal running *R* is "detected" by statistical methods, considering real process failures and related events.

In Chap. 5, which is dedicated to working means technical management, we will approach equipment reliability and maintainability matters; we will also discuss criteria for minimizing unavailability times that influence real productive capacity.

3.3 Working Time Analysis Methodologies

Let us describe working time analysis techniques generally applied in manufacturing industries:

(a) Stop-Watch Study Analysis
This technique was introduced in early 1900 by F.W. Taylor in the so-called *scientific organizations of works* and was later successfully extended to other industrial compartments for series productions. Starting in 1970, it was strongly criticized from a social and political point of view, effectively lessening its employment. Recently, it has been newly applied, thanks to some methodological innovations introduced.

For measuring time, it is necessary to proceed as follows:

(1) Workers are trained to apply working standards preventively and are informed about ways to proceed garnered from stop-watch study analysis;
(2) Through statistical criteria, the number of operative cycles to observe is determined, also taking into consideration dispersion phenomena of the analysed process if the dispersion is high, more samples will be needed for an effective analysis;
(3) For each macro-phase of the operative cycle, the time employed is recorded;
(4) Each working speed observed is compared to the "normal" speed, so that eventual adjustments can be determined;
(5) Time adjustments (increasing) for each working phase are applied according to specific tables, taking fatigue and physical efforts exerted during the working shift into consideration.

Applying this methodology is possible if the analyst is concerned about technologies of the analysed phases, ensuring in advance that working conditions will follow the manufacturing engineering plan, and proposes improvements that will become necessary. He must also observe movements taking into consideration ergonomics, defining normal and sustainable working speed. This evaluation also implies the application of specific rules related to physical effort, fatigue, working positions and movement repetition.

As a consequence, for each working phase analysed, a time increasing coefficient is applied, according to a standard table, elaborated on with statistical and scientific considerations by international rules institutions, in association with experts in ergonomics, physiology and working means.

In case of uncertainty of working speed estimation, the analyst can perform observations on different subjects to reach a "calibrated" evaluation. Furthermore, he can also use an available "database", compiled from previous experiences.

For each group of operations, it is important that the analyst separates active times from inactive times resulting from waiting for machine/equipment or movements between different working stations. Actions to be adopted are related to inactive phases that are not added value phases.

(b) Time studying with modern technologies
This technique applies the same criteria as the previous one, the only difference being that time observations are performed through video recordings, appropriately positioned and remote controlled, allowing for the simultaneous measuring of more operations related each other from different points of view.

Using specific software, by watching these video clips it is also possible to emulate normal working speed and introduce the right adjustments on measured times.

Like the previous example, this technique requires informing the observed workers, and pointing out the targets of the analysis so as to obtain their collaboration.

The advantages of the stop-watch study technique are:

- an opportunity to observe simultaneous correlated activities performed by more workers;
- an opportunity to watch and discuss the records with the workers, involving them in the improvement of manufacturing engineering;
- precision in observing operations deployed in multiple phases, with long times, as process auxiliary operations (changeover, set-up times, preventive maintenance interventions…);
- the analyst's continuous presence is not necessary during video recording;
- an opportunity for analysis of the video clips by modern specialists (manufacturing engineering technicians, ergonomists, safety and environment experts…).

This technique is widely applied in Japan and has even begun to spread through other industrial realities.

(c) Analysis with pre-determined standard times
Methods Time Measurement technique (MTM) began in the first half of the last century, spreading out through industrial manufacturing compartments, first in the United States and then in Europe. It consists of preventive analysis that allows quantifying the time necessary for each working micro-phase included in the operative cycle. Measurements are performed in office and are very precise and objective, because they are based on a wide range of elementary statistical data (standard micro-phases of manual work). Application requires significant effort in terms of analyst technicians, with high cost and long elaboration times. For this reason, the MTM technique fits with a widespread number of repetitive operations.

Other simplified techniques have been derived from MTM, integrating more MTM micro-phases, such as MTM-2 (still used in North America), UAS, MEK, TMC and, more recently, WF, developed in Europe and adopted by different manufacturing industries. The trend is to uniform all of these techniques through ISO.

Pre-determination of working times requires a detailed analysis of operative methodologies, defining the operation time-sequence diagrams in the designing phase of the project, while also taking the positions of equipment, machines, tools and each material handled during production into consideration. So it is necessary to design the technological layout, defining exact working conditions and action range for the workers, according to what was defined in Sect. 3.1.

Pushed by the European Community, recent new criteria for risks resulting from working efforts, working positions and movement repetitiveness during the working shift (OCRA evaluation method) have been defined. To satisfy the rules, and improve productivity at the same time, it is very important to apply virtual analysis techniques, with particular attention to the most critical operations, so as to optimize workplace ergonomics and choose the best-fitting partner tools. ERGO-UAS and EAWS, recently refined, are the most known standard applications for time measurement that combine ergonomics standards to MTM predetermination of working times techniques.

The analysis of working means and material handling containers is also very relevant for manual picking operations.

It is also important to adopt, during the development phase, product and process solutions convenient for assembling operations, with the target of reducing physical effort and avoiding human errors (design for assembly).

(d) Instantaneous Observations Analysis

Let us discuss this analysis technique, taking into consideration that it is only fitting for auxiliary operations, including office operations.

It is necessary that the analyst have a thorough understanding of the functions and roles of the observed employees. He performs periodic inspections in the relevant areas and by a route established by statistical method, so as to cover all workplaces. During inspections, he must instantaneously evaluate:

- if workers are at their designated workplaces;
- if workers are active;
- type of operation performed;
- type of tools used.

In collaboration with the foremen, the analysts classify activities by:

- types of activity, in relation to operative role assigned;
- utilization of technical support (working means, ICT supports…);
- time required for meetings, consultations or other activities not strictly necessary for the operative roles.

Processing data observed, by statistical method, it is possible to define roughly:

- medium degree of activity;
- activities composed by typology;
- time necessary for meetings and consultations;
- degree of utilization of technical support.

Consequent to this analysis, foremen have the necessary elements for deciding:

- the refinement of assigned tasks;
- improvement of working means and workplace organization;

- improvement of meetings and consultation management;
- the adequate amount of staff to assure service level and reduce inactive time.

The results of these instantaneous analyses can be directly adopted by department managers, even in a non-analytical way, to improve organization and the balance of work. For this purpose, it is also important to analyse interactions between different departments operating in the same organizational process.

3.4 Man–Machine Interaction and Standard Working Time Definition

For the effective development of the manufacturing engineering plan and the work analysis study, it is necessary to focus on the operative cycle and on interaction between worker and machine.

Manual working phases for a machine operator are:

- stopped machine working phase (SM);
- running machine working phase (RM);
- periodic auxiliary operations, in which part is performed with the machine stopped and part with the machine running (they affect workers' activity in relation to intervention frequency).

By the operative cycle analysis, it is possible to define the workers' active time (AT) and then the eventual idle time (IT) caused by waiting for the completion of the operative phases of the machines.

IT can be reduced by assigning other tasks to the workers, on the condition that said tasks will be compatible with their operative process cycle time. We define "machine matching" as the sequential intervention of the same worker in regard to two or more operative units in the same technological area.

We define $ST_{i,n}$ (typically measured in hundreds of hours) as the direct manpower standard time necessary to complete a single operation of a specific productive cycle (i), relative to the product series (n) (see Fig. 3.2).

$\mathbf{ST_{i,n} = (AT + IT)_{i,n}/k}$ **= Operation Standard Time**, where:

AT is the addition of individual active time of workers on operation (i), according to the operation time-sequence diagram

IT is the addition of individual idle time

k is the number of elements (n) produced for each operative cycle.

Operative cycle analysis is particularly complicated when operations are performed in operative interconnected working centres, with simultaneous employment of more workers and alternation of automatic and manual operations in the

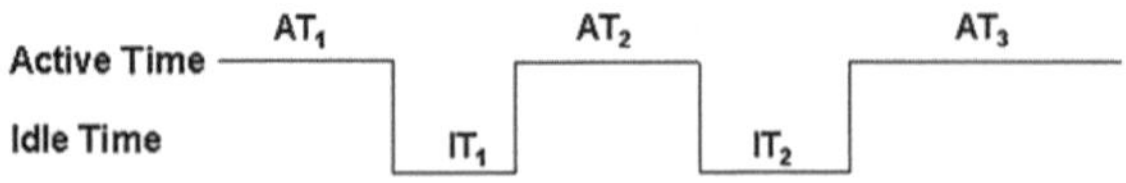

Fig. 3.2 Active time and idle time example

same "integrated system". For those "systems" with a high automation level, it is important to estimate manpower requirements for equipment management, technical checks and *periodic auxiliary operations.* In these situations, it is defined as the necessary "operative garrison", by applying work analysis techniques mentioned in the previous sections. We define $ST_{m,n}$ as the working standard time necessary to manufacture the product unit (n), in the "system" (m):

$\mathbf{ST_{m,n} = LH_m/SPC_n}$ = **System Standard Time** (labour hours for product unit), where:

LH_m are labour hours necessary for the operative garrison of the "system", during considered working time;

SPC_n is standard productive capacity relative to product series n, referring to the considered working time.

We define "Standard Time" ST_n as labour hours strictly necessary for manufacturing a specific product series n, based on operative cycles assigned.

ST_n is calculated adding $ST_{I,n}$ and $ST_{m,n}$ according to the bill of material, as seen in Chap. 2.

ST_n is the fundamental input data necessary for:

- **planning and assigning direct labour;**
- **checking direct labour efficiency;**
- **accounting direct labour cost for single products.**

ST_n is called *standard working time*, without technical and managerial losses. In Chap. 4, we will examine this argument again from the organizational point of view of labour.

Let us close this section by underlining the importance of interaction between labour activities related to the productive process and the ones performed by machines and equipment, according to modern criteria of labour organization.

Automation can eliminate repetitive and tiresome manual jobs. When not strictly tied to the cycle time of productive machines, workers can perform other supervisory tasks such as quality checks, tool changes, process parameter adjustments and recurrent preventive maintenance operations (when a specific technical competence is not required). For more complex operations, it is useful to provide modern monitoring techniques for "guided conduct" and for keeping an eye on critical process parameters even from a distance.

Automation is really an advantage when it improves productivity and process quality. Avoiding working conditions that cause fatigue and increase the risk of human errors, the processes become more available, with better product quality and more stable production flow.

Another intelligent way to improve labour productivity by reducing ergonomics' issues in the workplace is the usage of the so called "Low Cost Automation" (LCA); it consists with the usage of intelligent automatic systems that mostly utilise mechanics/pneumatic solutions combined with gravity force to reduce the human efforts in loading and unloading parts on the

equipment and in handling components and raw materials. This solutions are cheaper than hard automation solutions and allow a rapid return of the investment.

A higher automation level typically requires a higher investment, for which economic payback can be obtained by variable cost reduction of labour, under the condition that technical systems can be used for a long life-cycle and with a high degree of utilization. For this reason, the matter of equipment and machine utilization is broadly examined in this section. Before we approach this argument, it may be suitable to reread the consideration in Sect. 1.7 for strategic system characteristics.

3.5 Equipment Utilization Analysis

In the following sketch, the deployment of working times is represented, starting from what is possible for defining Key Performance Indicators (KPIs) of systems (see Fig. 3.3).

Assuming that the manufacturing system is made of synchronized and interconnected operative units, working according to standard operation sequences, **system KPIs** are:

WCT (in minutes)

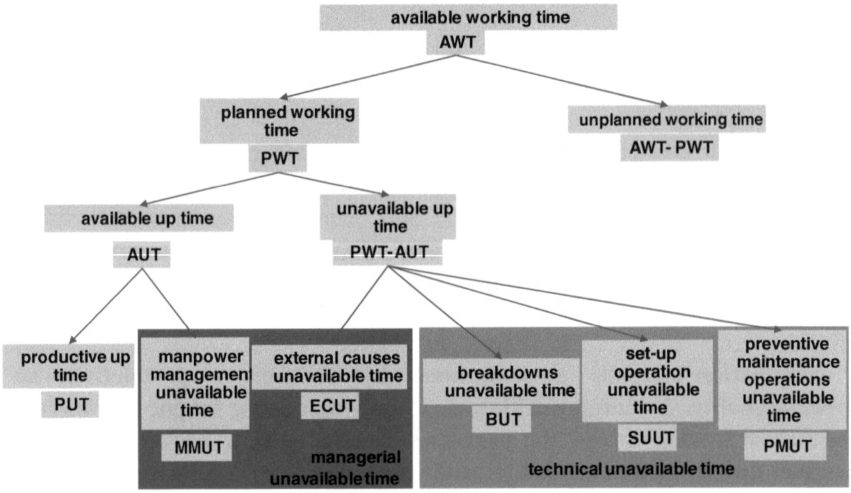

Fig. 3.3 Equipment working times deployment diagram

	= **working cycle time**, which determines the system's productive cycle time;
PUT (in hours) = WCT • QP/60• k	= **productive up time**, referring to working time employed for production, QP being the quantity of elements produced;
R = PUT/(PUT + BUT)	= **technical reliability;**
A = AUT/(AUT + BUT + SUUT + PMUT)	= **technical availability;**
OEE = PUT/EWT	= **overall equipment efficiency** (or global equipment effectiveness), EWT being the employed working time to produce quantity QP;
U = PWT/AWT	= **utilization degree** of production capacity available (equipment saturation).

In case of N equivalent modules set in parallel, WCT and R are relative to single modules, while total system productive capacity is N·APC (Available Productive Capacity for the single modules). In case of N modules directly connected in series, $R = R_1 \cdot R_2 \cdot \ldots R_N$, R_n being the availability coefficient of a single module.

For an integrated system", measurement of Overall Equipment Efficiency, for series production, can be obtained from the following calculation:

$OEE = \frac{\sum_{n=1}^{N} \frac{QP_n}{HVP_n}}{WTE}$ = **Overall Equipment Efficiency**, where:

N is the typology of transformed product (element code);
QP_n is the quantity of element (n) produced during the period WTE, minus unrecoverable scraps;
HVP_n is hourly virtual production relative to the single element n;
WTE is the working time employed, measured in hours.

The quantity of elements produced QP_n is measured from the *control system.*

The difference (A-OEE) is due to inefficiencies in the operative system management, for the following six potential causes of loss:

(1) **lack of parts** (materials) **in input or** (final product) **in output** in the "system";
(2) **no availability of auxiliary materials or power supply;**
(3) **changeover out of production plan;**
(4) **no re-workable scraps of production;**
(5) **start-up and shut-down losses**, due to learning curve of workers and technical try-out of the systems;
(6) **speed reduction** for lack of workers, low labour efficiency, lower than standards tool life…

3.6 Introduction of Integrated Production Systems

At this point, it is important to examine "integrated production systems", which are hardly employed in the automotive industry.

An "integrated system" is made of more process stages or sub-processes, directly connected to each other by a one-piece-flow concept, able to run complete manufacturing cycles and governed by automatic control systems interconnected by internal nets. Each of these sub-processes could also be a single "module" (tool machine, technological equipment, machining centre, assembling cell or line part), or two or more "modules" working in parallel. We remind you that these "modules" also include quality control stations.

In Sect. 3.1, we examined the technological layout study, which defines the links between operative units and points out flow and stocking areas for process materials. The analysis of the "time sequence diagram" of each operation leads to identification of the "leading stage", which determines the operative cycle time of the integrated productive system (bottle neck concept).

During the setting phase, it is important to balance and size up the different process stages, so that they may have the same productive capacity. Nevertheless, ideal balance is not always possible and it is fairly normal to set "bottle necks". Then, shop floor management must focus its attention on the "leading stage" presenting the minor Standard Productive Capacity (SPC), searching for:

- optimization of time-sequence diagrams (minimum WCT maintainable);
- auxiliary operations, affecting optimization of idle time (minimum *pu* possible, considering the same WCT);
- standard application of preventive maintenance, for improving technical availability (maximum *A* obtainable).

In searching for optimal conditions in operations between modules of the balanced integrated system, it is important to use, in the development phase, modern CAPE supports, which allow for the simulation of functioning conditions, with particular reference to production flow and inter-operational buffers. For effective analysis, it is necessary to consider the following parameters relative to the single "operative unit" in the software: WCT, duration and frequency of technical stoppages consequent to auxiliary operations and breakdowns (MTBF, MTTR, according to definitions reported in Chap. 5), position and dimension of "buffers".

Let us conclude the argument for manufacturing "integrated systems", emphasizing that they are equipped with a 1st level information technology net, which links machine control units and automatic devices (CNC, PLC…) to microchips dedicated to data collection and processing.

This 1st level information network works for operative cycle planning, production flow and monitoring of auxiliary operations (tool change, maintenance operations, quality statistic control…). It can be linked to information systems at a superior level, dedicated to the overall management of production planning, to statistical analysis of quality data, or to the traceability of components and operations related to product quality.

For complex "systems", with a high automation level, it is important that this information technology collect and transmit net data to a dedicated information technology centre, distinguishing events caused by technical failures from those resulting from the normal auxiliary operations, rather than caused by machine users. In this way, data can be processed in time to compliment "reports" of "system" efficiency (technical and overall), making operative control easy for responsible employees.

The concept of an "integrated productive unit" is consequent to the attribution of management responsibility to a single manager, who has the collaboration of an "interfunctional team" made of experts and the support of information technologies. Work organization applies, as much as possible, to "job enlargement" criteria; for example, assembly line workers are assigned quality self-inspection tasks, machine operators given autonomous inspection and maintenance tasks, and so on.

3.7 Operative Productivity and Flexibility

Productivity of a manufacturing system is evaluated basing on the following "functioning parameters", according to the definitions mentioned earlier:

HVP = hourly virtual production
A = technical availability
APC = available productive capacity, referring to daily available functioning time
ST = standard time referring to the transformed product unit.

Operative flexibility of a manufacturing system is evaluated based on the following process and product characteristics:

- element range producible at "random", that is to say without interrupting the process continuity;
- element range producible by batches, by tool changes requiring specific "set-up time";
- product mix re-ordering frequency, respecting available productive capacity and considering the above-mentioned "set-up-times", applying the criteria of minimum sustainable batch (mentioned in the Basic Concept Chapter);
- medium lead time (the meaning of which will be explained in Chap. 6).

When "benchmarking" research is made for manufacturing systems, it is important to compare parameters of both productivity and operative flexibility.

3.8 Equipment and Machine Loading

Let us now discuss the planning of operations and productive capacities, very important for the success of industrial activities, especially when significant investments are involved. For automotive applications, we generally have to consider monthly plans that cover a period of one or more years. Working planned time is expressed in hours/day and in working days/month. Necessary input data for equipment and machine loading are:

- product typologies "N" assigned to the "system";
- monthly production plan (MP_n), related to the specific product n;
- virtual hourly productivity (HVP_n);
- standard productive capacity (SPC_n), intended as a medium data for one hour of working time;
- OEE (overall equipment efficiency);
- Available working days in a month (MD).

Planned working time calculation (machine loading time) can be done applying the following alternative criteria:

1st CRITERIA is referred to as standard productive capacity (SPC), without considering any lost time for process failures (breakdowns and speed reductions) and management inefficiencies.

NPWT $= \sum_{n=1}^{N} \frac{MPn}{SPCn} / MD$ = **net planned working time (hours/day)**.

This criterion considers that the total available working time (AWT) needed to recover eventual production losses, due to conduction or technical failures, will widen working time when necessary. Alternatively, it is necessary to set safety stocks.

This criterion is normally applied to machine loading, when we have autonomous machines, or multi-operational systems not strictly interconnected, with the possibility of working in a warehouse.

In the case of more than one operative unit, or module, operating parallel to the same department, NPWT is calculated as medium unitary data.

This is the best way to plan machine/equipment loading used for steel printing, polymer printing, mechanical machining centres and electrical and electronic components production.

2nd CRITERIA is referred to as the virtual productivity (HVP) and considers the Overall Equipment Efficiency (OEE).

PWT $= \sum_{n=1}^{N} \frac{MPn}{HVPn} / \mathrm{OEE} \cdot \mathrm{MD}$ = **planned working time (hours/day)**.

This criterion is applied usefully to *multi-operational systems*, organized in multiple phases and interconnected to each other with a one-piece-flow concept. Its application requires periodic statistical utilization analysis for the systems, using the criteria from Sect. 3.5. This mode of work is useful for the following automotive productions:

- body assembly and painting for cars and light commercial vehicles;
- space frame assembly and painting for industrial vehicles;
- engine and transmission assembly;
- electrical device assembly;
- foundries for iron cast and aluminium elements;
- steel fusion and lamination, with an integrated cycle;
- tire manufacturing, with an integrated cycle;
- glass parts manufacturing, with an integrated cycle.

Degree of utilization of planned productive capacity, also called "equipment saturation", is measured by:

u = **PWT/AWT,** where:

AWT is the total available working time (hours/day), considering duration of the work shift and the number of shifts available in a week, assuming one has all necessary labour available.

For automotive production activities, a monthly analysis of machine loading (PWT) is normally performed, considering a horizon period of at least 18 months.

This planning process, through specific temporal diagrams, focuses on the following data:

1. virtual productive capacity evolution (HVP) of the system, which can be variable in time depending on structural empowering or re-dimensioning;
2. evolution of overall equipment efficiency (OEE), depending on improvement plans;
3. machine/equipment loading forecast trend (PWT), expressed in hours/day and in weekly working shifts, calculated based on the production plan and considering the above parameters for effective productive capacity;
4. available productive capacity utilization degree forecast (U), in the time period considered.

It is important to note that the above productive capacity planning criteria allows for the fulfilment of all programs, assuring the most convenient distribution of the working load among the different productive units/plants, in consideration of the available resources. In this way, the daily and weekly working shifts are planned for the run of the production plan.

Whenever critical situations affecting the degree of utilization arise, it is necessary to adopt tactical solutions (temporal re-organization of production scheduling programs between the plants, longer or shorter working times compared to the norm) or strategic solutions (structural empowering or re-dimensioning of available productive capacity).

These decisions are made depending on medium/long term commercial planning and considering the risks of over-saturating the productive capacities planned. In case of significant decreasing of the saturation, production costs will worsen, due to a minor absorption of the fixed cost of the structure. In the case of significant over-saturation, delays in delivery of the final product to the customers will occur, with probable market share losses and profits decreasing compared to theoretical forecasts.

These considerations underline the importance of correctly setting productive capacities, in relation to market opportunities.

3.9 Defining Installed Productive Capacity

To conclude this chapter, we are going to connect the *manufacturing engineering plan* discussed in Sect. 3.1 and the implementation of the machine loading criteria just discussed in the previous section.

Let us remember that productive capacity is first planned for each industrial site in relation to market targets (specific product annual sales plans): During the product's industrial life cycle, it is typical for there to be a peak of sales demand, related to the year of greatest success in the market.

Let us define QY_m as the yearly demand forecast during the life cycle and QY_p as the yearly demand in the peak year. If project productive capacity is set in relation to QY_p, the medium utilization degree of productive capacity will be: $U = QY_m/QY_p$.

Beneath the threshold value of U = 80 %, for "capital intensive" productions, the incidence of structural fixed costs can cause a critical situation for the enterprise's earned income. So, in planning productive capacities, Top Management must research a balance between requests from Marketing and investments required by Production Management, calibrating industrial initiatives in relation to real market possibilities and profitability targets.

It is clear that a manufacturing system with a bigger operative flexibility and convertibility for new product insertions has a low risk of being less used due to a specific product's life cycle. Additionally, a Component Maker is more likely to have the right degree of utilization of its productive capacity, when manufacturing systems are multi-functional and suitable for supplying more industrial customers, reducing market risk.

Considering that, it is possible to use the following two alternative criteria for setting "manufacturing systems".

Applying the *1st criteria* for machine loading calculation, Available Productive Capacity (APC) relative to each similar group of machines or autonomous technological modules (machining centre, multi-station or single press, machining cell…) is set, assuming that solutions implemented will correspond to the technical–economic criteria mentioned in Sect. 1.4. Consequently, machine tools and technologically similar modules operating on parallel requirements are calculated by:

MR = **machine requirements** (number of modules required) = **QY/(APC • AWT)**, where:

(QY) is the quantity of product required yearly, based on the marketing plan (including spare parts required for after sales);

(APC) is the available productive capacity (medium data related to each module);

(AWT) is the maximum available working time for production in a year, expressed in hours.

MR should theoretically be counted to the next full number, if it is not possible to use overtime to balance eventual marginal lacks, or partial *out-sourcing*.

For "one piece flow integrated productive systems", discussed in Sect. 3.6, the **2nd criteria to calculate** machine load is applied. Productive flow speed necessary to satisfy market demand is determined by:

$HVP = \frac{QY}{OEE \bullet AWT}$ = **Hourly Virtual Productivity** required by project, where:

(QY) is the quantity of product required yearly, based on the marketing plan (including spare parts required for after sales);

(OEE) is the overall equipment efficiency foreseen by the project for the manufacturing system;

(AWT) is the maximum available working time for production in a year, expressed in hours.

It is to be considered that, for each technological process, productive flow speed is characterized by a maximum sustainable threshold level. In fact, beneath a certain level of Working Cycle Time (WCT), inactive phases hardly influence the time sequence diagram. When this occurs, it is good to share productive flow between two or more "parallel systems", searching for the best conditions, as much from an investment point of view as a transformation cost point of view.

Both criteria for productive capacity dimensioning are based on annual product sales plans, during their industrial cycle life, without considering the effects of market season trends. It is assumed that eventual monthly requirement peaks can be sustained by longer working shifts and extra working time during the year. It is also possible to produce in advance, but only if there are guarantees of market absorption during the period of peak demand. If all this should not be enough, longer lead time on delivering is to be considered, by managing order reservations.

Value MR and HVP determine productive capacity by project and are the input data necessary for the manufacturing engineering plan and for the executive project for "manufacturing systems", following the methodology discussed in Sect. 3.1.

Starting from the definition of Hourly Virtual Productivity (HVP) above, it is possible to transpose the concept of **Takt Time, meaning the ideal pace of a production system for satisfying the customer demand:**

$$TT = \frac{1}{HVP}$$

The Takt Time is not the rhythm at which the customer is asking for products but is representative of the ideal pace the production system must have in relation to the average customer demand in the period considered.

Setting the production system to the Takt Time requires some preliminary conditions, such as:

- stability of production system (high technical availability and efficiency)
- levelling of customer demand in the period considered.

Once the Takt Time has been set, it is more effective to maintain it in as stable a condition as possible to consolidate production system standards and obtain the best level of performance.

Chapter 4
Work Analysis and Labour Productivity Evaluation Criteria

4.1 Activity Level and Labour Efficiency

By *direct labour*, we mean the personnel directly involved in the product transformation. In mass production, the activity levels are measured in standard working hours necessary to complete the production plan. In fact, considering the same cost factors, the added value applied to the product can be considered proportional to the productive hours performed.

Let us see how to plan and check the operative efficiency of direct manpower in mass productions.

In Chap. 3, we have seen how to determine the standard working time $(ST)_n$. It corresponds to the hours of direct labour necessary to transform the product series *n,* complying with the relative *manufacturing working cycle*. We remember that the specific ST_n are assigned to the single product codes and aggregated according to bill of material, depending on the "make/buy" levels (as seen in Sect. 2.3).

Beginning with the assigned program and production progress, the shop floor information system assigns to each productive unit (corresponding to an industrial responsibility centre) the Planned Activity Volume (PAV), or the Developed Activity Volume (DAV), both measured in production "standard working hours". The one that generates "sales" is considered the main production, being assigned to the market of the enterprise and its own "business units".

$$\mathbf{PAV} = \sum_{n=1}^{N} QA_n \cdot ST_n = \textbf{Planned Activity Volume},$$

where:

QA_n is the quantity of products *n* assigned to the plant or to the shop unit, according to the production planning and referring to a specific working period (working days, weeks or years);

ST_n is the working time determined by the standard operating procedures, measured in working hours for each one of the product series (n);

N are the series of product charged to the productive unit;

M. Gobetto, *Operations Management in Automotive Industries*,
Springer Series in Advanced Manufacturing, DOI: 10.1007/978-94-007-7593-0_4,

PAV is the number of standard working hours necessary to perform the assigned plan of production This evaluation criteria allow for the merging of data belonging to different departments and to different manufactured products.

4.1.1 Labour Efficiency Analysis

Let us define HDL as the total amount of direct labour working hours collected by the information technology system in the productive units.

Let us also refer to DAV, in the considered period of time, measured in standard working hours and quantified according to evaluation of the progress of production.

The direct labour efficiency is measured by the output/input ratio: $\mathbf{DAV/HDL} = \boldsymbol{\eta}$.

This statistical indicator represents the level of utilization of direct manpower, from a productive point of view.

The difference (HDL–DAV) corresponds to the time in which labour is on the working site without contributing to the main production plan. The causes of this loss can be divided as follows:

(1) inactivity due to excess of available production capacity during the working shift or to excess workers assigned against actual requirement;
(2) time lost for external causes in the productive unit (lack of necessary incoming materials, lack of withdrawal of out-going products, lack of energy…);
(3) time lost for technical breakdowns of the "systems";
(4) time necessary for extra-standard operations, such as solving "non-quality" or "technical failure" problems;
(5) time lost for rejects or reworking;
(6) other causes (operations performed in lengthier time than standard, speed losses in the process, training of workers…).

To assure the reliability of data related to the above eventualities, it is important to have a rigourous control systems. For example:

- hours lost to cause (1) must be dealt with by Operation Planning and Human Resources Departments;
- hours loss to causes (2) and (3) can be collected automatically, using the information technology systems, and must be dealt with, respectively, by the Material Supply and Maintenance Departments;
- rejects and re-workings of cause (5) must be checked by the Quality Assurance Department;

- extra-standard operations related to cause (4) must be dealt with by the Manufacturing Engineering Department;
- the remaining amount of working time not used for production (cause 6) is normally assigned (as per difference) to inefficiency and also includes the effect of the so-called "learning curve", during new product launches.

To achieve the targets in terms of product quantity and costs, it is necessary to check labour efficiency (η) monthly at the very least, analyzing the main causes that determine the deviations. The plant information technology system is normally set to elaborate this data automatically for each one of the productive units and in time to make corrective actions possible.

4.2 Manpower Planning

Estimating the manpower requirement for production means considering local legal constraints for working times (different from region to region and depending also on agreements with local unions), as well as statistical records on labour present on the job. The substantial input data are:

- working days available for each worker during the considered period of time, minus collective holidays;
- duration of daily working shifts, minus free time for lunch;
- shift organization and rotation;
- sick leave, accidents and incidents of permission for individual leave.

Direct Labour Requirement (DLR) can be obtained considering the planned activity volume (PAV), compared to the individual amount of activity achievable, both being measured in "standard working time", referring to a specific period of working time (day or week).

DLR $=$ **PAV/IAA**, where:

PAV corresponds to the calculation in Sect. 4.1.

IAA $= \eta \cdot IWH \cdot$ **(1 − a) = Individual Activity Achievable**, equal to the productive hours worked on average by each worker, being:

IWH individual working hours, corresponding to the hours of presence on the job determined by the collective agreement during the working period considered (average data that also considers shift organization);

η direct labour efficiency (measured as indicated in Sect. 4.1);

a absenteeism index, estimated on statistical data;

DLR direct labour requirement, in other words, the number of workers necessary to carry out the assigned production plan, complying with the working hours, with the average labour efficiency and also considering predictable absenteeism on the job.

Normally PAV and IAA refer to a single working day of planned shifts (according to the PWT calculation reported in Sect. 3.8). Conversely, when there are rest shifts on rotation during the week to be considered, it is necessary for PAV and IAA to refer to the complete working week.

These criteria can be simplified, assuming the direct labour productivity as a standard value, measured by the ratio QP/DLR, where QP is the produced quantity. This criteria does not consider the variation due to product mix and the evolution of other productivity factors mentioned above (ST, η, a).

4.2.1 Indirect Labour Requirement

In addition to direct labour, there are other workers necessary to production management that must be considered in the manpower requirement; these workers are connected to auxiliary activities that are not specifically included in the standard operative procedures (indirect labour).

The principle auxiliary activities that contribute to the indirect labour requirement are:

(1) **Quality Checks** "off line" (metrology rooms, quality labs, statistical control), the requirement of which is established by considering the quality control plans and checking methodologies; this requirement depends both on the PAV and on the number of planned working shifts, because a minimum fixed number of workers is required for each shift;
(2) **Ordinary Maintenance of Equipment and Tools**, the requirement of which is estimated considering a statistical work load and the required service level; it depends both on PAV and on the number of planned working shifts;
(3) **Material Handling** on the shop floor (in addition to what is included in the standard operative procedures), the requirement of which depends on the means and methodologies of handling, both on PAV and on the number of planned working shifts;
(4) **Warehouses and Dispatch Management**, the requirement of which is principally related to the number of planned working shifts, the workers being employed for fixed garrisons;
(5) **General Plant Services**, not directly connected to productive processes and the requirement of which are determined by fixed garrisons during the week, for industrial supervision and general equipment conduction.

In general, functions (2), (3), (4) and (5) are partially "out-sourced" to specialized providers; nevertheless, the manpower requirements should always consider the above-mentioned criteria.

In the same productive unit or in an entire plant, considering the same arrangement of services outsourced, **the incidence of indirect labour** represents a necessary statistical data for evaluating labour efficiency:

i = HIL/DAV, where:

HIL is the total amount of indirect labour working hours;

DAV is the developed activity volume, in the considered period of time and in the same productive unit, measured in standard working hours and quantified according to evaluation of the progress of production.

It is important to consider that a part of the auxiliary activities mentioned above (not included in the standard manufacturing activities) can be assigned to direct labour; in this case, these hours should be included in HIL.

For the same manufacturing system and for the same volume of activity PAV, with some organizational improvements, it is possible to minimize the incidence (i).

Other activities, such as die construction, prototyping, laboratory testing and product experiments, are considered R&D activities and, for this reason, not related directly to the management of manufacturing.

4.2.2 Operative Plan and Staff Balancing: Requirements

Comparing the available manpower staff (MS) with the labour requirement (LR), including both direct and indirect labour, it is possible to evaluate excesses or deficiencies in the staff for each productive unit or plant, carefully distinguishing any specialization that could be a constraint. To assess the likely evolution of the staff during production, without anticipating new hires, it is necessary to consider the normal ratio of outgoing personnel minus any necessary substitution.

The dynamic difference between the available staff and labour requirement is measured by % variation:

$$D\% = (LS - LR)/LS$$

where:

- LR is the Labour Requiremt, made of direct labour and indirect labour.
- LS is the Labour Staff at disposal, counting direct labour and indirect labour.

If the indicator is positive, it means that there is an excess in the staff; if it is negative, it means that there is a deficiency in the staff. The initial balance between staff and requirement should be worked out at the plant level, moving the workers among the shop floors to minimize inefficiencies in the working hours and in the production flow.

If D % shows *low staff deficiency*, it is possible to make up for it through overtime, as per local standards and agreements (different country by country) or the use of temporary workers.

If D % shows a *structural staff deficiency*, at a stable or growing trend, the first countermeasure is to cover the turn-over and then immediately hire new workers, even if it is not possible to transfer part of the production plan to other plants.

If D % shows *temporary excess in the staff*—according to the medium term operative plan—it is possible to reduce the monthly working hours proportionally, managing compensations during the whole year. This countermeasure is recommended to compensate for the seasonal oscillations in the sales or *"phase-out/ phase-in"* situation related to new product industrialization.

If D % shows *structural excess in the staff*, at a stable or growing trend—according to the long-medium term operative plan—it is necessary to take extraordinary action such as:

- assignment of new additional productive programs to the plant, complying with the general enterprise productive plans and after a feasibility check (making sure productive capacity is available and looking into the eventual necessity of additional investments);
- reduction of working days in a month/year, or temporary staff reduction through flexibility tools (temporary suspension from the job, example: CIG in Italy; reduction of working shift…);
- definitive reduction of staff through the use of mobility plans toward the external, although only when the above-mentioned countermeasures are insufficient.

Considering the above, it is clear that achieving a balance of productive and sales capacity is the real key, especially because of the uncertain nature of the automotive industry, with demand from the market being very variable and individual economic and political scenarios not easily predictable.

An operative planning process, as described in detail in Chap. 6, is an attempt to equilibrate the monthly production plan with sales demand, considering the productive capacity installed and the working staff available. Safety buffers are the final and intermediate product stocks; however, these stocks should be limited to the most requested models and versions, not determined by specific optional contents; in addition, their dimension should be limited as much as possible to physiological levels, to allow short lead time in delivering to the customers and avoid overstocks that increase the WIP level and influence the sales policies.

Bringing to mind what was discussed in Sect. 3.8, we would also like to remind the reader that the manpower requirement setting should be calibrated according to the "technical system" productive capacities, searching for the more convenient working shift rotation over the course of a week.

4.3 Working Time Length and Flexibility

Referring to Chap. 3, let us first remember that, with the same hourly virtual production (HVP), the weekly and annual productive capacity grows depending on the following parameters:

- hours available for the plant to be functioning (AWT), in regard to local and/or national laws and working hours for personnel set by union agreements;

- overall equipment efficiency (OEE), in relation to their technical availability (A) and to management of productive processes.

To answer qualitative variations (model mix) in customer demand quickly, it is fundamental to have the availability of **flexible and programmable systems** (see also Sect. 1.7).

To answer quantitative variations (total product volume) in customer demand quickly, it is necessary to have a certain **flexibility in the working hour**, to adapt and refine the weekly activity levels, complying with the monthly operative plan.

To understand this last concept, we may examine the approach of Japanese industries to working hours; they are characterised by a very high cost of work per unit, very high productivity and the promise of stability for the working staff over the years.

Japanese industries normally do not plan their activity by covering the twenty-four hours available in a day with three shifts per day for 5 days a week, as many western enterprises do to utilize their equipment and investments at the maximum level; the Japanese tend to set their production with two daily shifts for each of six days a week; these two shifts are opportunely spaced, so that they can be extended or shortened (within a certain amount of time) on demand and by necessity.

A consistent number of breaks between shifts guarantees the possibility of performance of the majority of auxiliary operations requiring stoppage of the machine (die changes, time-based maintenance operations…); in this way it is possible to improve the OEE during the planned working time. The working hours of maintenance workers are consequently adapted to that of their production counterparts.

It is possible to have a daily minimum of 14 h to a maximum of 18 h of full equipment running (corresponding respectively to seven or nine h of working time per shift). In this way, with the same staff, the utilization of productive capacity can be stretched equal to 25 % and thus allow the activity level to fit with the monthly market and sales trend.

In other social contexts (especially in Western Europe), it is practically impossible to apply the "Japanese way", or at least as well as they do. Due to laws and terms of union contracts, extensive collective overtime is simply not possible. It is, however, important to find different agreements complying with local laws to fit with the concept of flexibility.

In conclusion, for a *capital intensive* enterprise to stay competitive, having the possibility of extending the available working time in a week (AWT) to six working days is fundamental, utilizing maximum productive capacity when customer demand requires. At the same time, it is also essential to have a stretched daily/weekly working time to adapt the activity level to market needs, even if this should imply the following:

- rest times during the week on rotation, to respect the working hour limits set by law or union agreement;
- appropriate relaxation spaces facilitating extended shifts;
- transfer services for the employees to fit with the extended shifts.

This argument is relevant not only in the automotive industry but also in other industrial compartments, especially for those that are subject to variable customer demand and occupational restrictions. It is also important to underline that, from a social point of view, higher flexibility in working time means stability in job occupation and better success of the enterprise on the market, increasing the chance of further development.

4.4 Labour Productivity and Improvement Plans

Labour productivity is related to individual productive capacity and can be calculated by analyzing the quantity produced in a specific period of time, such as a working year, divided by the dedicated labour workforce, that is, the **Individual annual Labour Productivity (ILP)**:

$$\mathbf{ILP} = \mathbf{QP}/\,(\mathbf{DL} + \mathbf{IL})$$

where:

- QP is the quantity of products produced in the year by plant (number of vehicles, engines, transmission, components…);
- (DL + IL) is the average value of the staff during the year, including direct and indirect labour.

To increase ILP, it may become necessary to decrease the workforce employed for manufacturing a specific quantity of product (see earlier direct and indirect labour requirement calculation), so ILP mostly depends on the following factors:

$\mathbf{ST_n}$ standard working time for each product series n
η direct labour efficiency
i indirect labour incidence

The reduction of standard working times ST_n is the first relevant factor for the improvement of productivity and the reduction of "transformation variable cost ". To reduce ST_n, it is necessary to analyse it with accuracy; it is mainly given by

$$ST_n = AT_n + IT_n$$

where AT is the active time assigned to operators and IT is the operator's idle time due to technical factors (i.e., machine downtime or other technical requirements) or unbalancing, which keep operators idle. So, the first improvement is the reduction or possible elimination of the idle time IT.

Improvement activity should then be concentrated on looking deeply at active time, which can be divided into two different categories:

$$AT_n = NVAA_n + VAA_n$$

where:

1. NVAA (Non Value Added Activities) are the activities that do not generate value for the final product (handling, moving, walking, turning, waiting…).
2. VAA (Value Added Activities) are the only activities that generate value for the final product (assembling, screwing, fixing, painting, welding…).

Reduction is possible by concentrating on the NVAA portions of the standard working time; some of them are harder to eliminate and can only be reduced, while others could be eliminated completely through the application of the following criteria during the design or improvement phases:

- workplace organization improvement activities (ergonomics improvement, application of visual management on standard procedures and instructions, optimization of material presentation…);
- continuous improvement of working tools, equipment and methods;
- application of low cost automation (LCA) as mentioned in Chap. 3 at the of Sect. 3.4 and/or hard automation for handling activities (remaining NVAA);
- product changes in terms of "design for manufacturing".

It is very interesting to note that the world class carmakers, inside their standard working time, are still far from 100 % of VAA, which indicates that big opportunities for continuous improvement are still in place. Of course, those values depend on the level of detail by which motion is studied: the more you deploy the single activities, the higher the opportunity to isolate and remove NVAAs will be.

Engineering activities in workplace organization have to address implementation of the following steps:

1. Analyse and reduce/eliminate ergonomic issues (MURI); those activities require specific knowledge of physical efforts and working conditions in order to minimize the impact on worker activities in the workplace; particularly relevant are the concepts of the "**golden zone**" and "**strike zone**" in which operators can move more easily by minimizing all physical effort, as shown in Fig. 4.1. As described in Table 4.1, each of the areas in the golden and strike zones are classified based on the difficulty in handling items (parts to be assembled and tools) in the working areas.
2. Analyse and reduce variation in the operation (MURA); those improvements can be achieved by analytically examining the sequences of the operations and strictly applying the principle of standard work, in order to minimize variations between different operators and those performed by the same operator Fig. 4.2.
3. Finally, after having reduced *MURI* and *MURA*, the remaining NVAA activities (*MUDA*) can be drastically reduced and eliminated by focusing on material handling operations affecting the operator's work, such as re-organizing the material's display on the side of the line and dealing with other unnecessary movements, possibly by use of some intelligent low-cost automation solutions. In Fig. 4.3, an example is shown of such an optimization on an assembly line by simply feeding the parts to the line by the right logistic method (further details

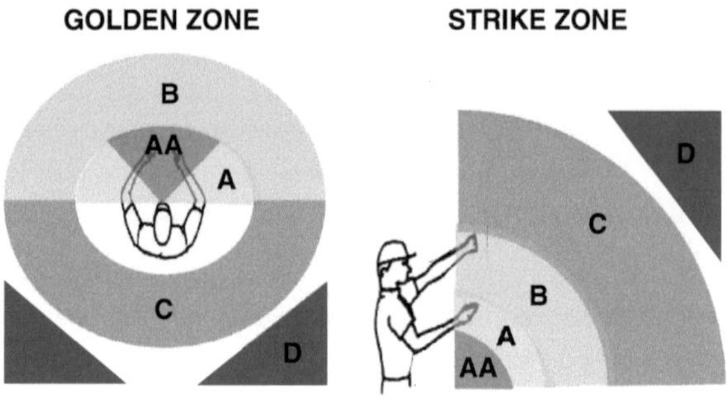

Fig. 4.1 Strike and golden zone concept

Table 4.1 Strike and golden zone areas classification

	Golden zone	Strikezone
AA	All the items can be provided to the assembly point within the field of vision and without changing the heights of the item supply points	All the items can be handled at the right height without raising hands
A	Items are placed within a region three times as big as the assembly unit. Items can be taken by stretching out elbows. Both hands can be used	Items can be handled at the right height but by raised hands
B	Items can be taken by stretching out elbows although they go up beyond shoulder height. Items are placed within a region six times as big as the assembly unit	Items can be handled by raised hands through extending arms and raising elbow over the shoulder
C	Items can be picked up by turning the body	Items can be handled only through considerable extension of body and arms and/or using additional step/ladder, etc.…
D	Items can be fetched by walking	Items are totally out of range and the worker needs specific aids to reach them

on Logistics and Supply Chain Management will be addressed in Chap. 6) and presenting necessary tools at the right spot, actions that drastically reduce the need for operators to walk away from their stations.

Coming back to increasing labour productivity in general, the **second important factor is labour efficiency** η, which proportionally influences productivity. As seen at the beginning of this Chapter in Sect. 4.1, labour efficiency is strongly influenced by process stability, so main keys to influence positively $\boldsymbol{\eta}$ are:

1. striving for process stability and high level of equipment efficiency (OEE) through the application of Autonomous Maintenance and Professional Maintenance methodologies to minimize failures due to a poor Maintenance (see Chap. 5);
2. stabilizing and synchronizing Internal and External Logistics to guarantee a continuous feeding to production lines (see Chap 6);

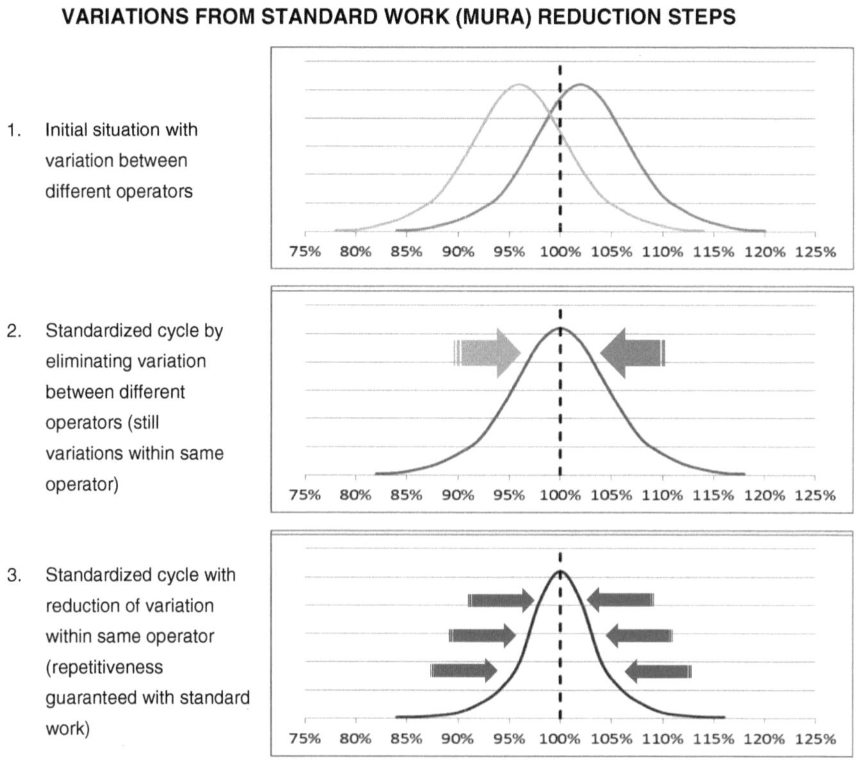

Fig. 4.2 Reduction of variations

3. searching for first time quality by engineering more robust processes; this is possible by working on labour intensive processes in minimizing human error possibilities through the application of "standard work" logic and labour effective training; conversely, for capital intensive processes it is necessary to reinforce the control of component and parameter that directly influence quality aspects; all these action relate to Total Quality Management methodology (see also Chap. 8).

The **third factor is containment of indirect labour incidence i**, because this labour factor does not contribute directly to the value of the final product. It is important to underline that all the above mentioned actions necessary to improve direct labour efficiency are also beneficial for a better organization of indirect labour force and its containment.

The utilization degree for labour corresponds to: $\boldsymbol{\eta}/(\mathbf{1}+\mathbf{i})$

This indicator, considering the same equipment and product, can be progressively incremented by acting on the labour.

For the automotive industry (Final Producers and Main Component Producers), labour still represents the main factor, even if automation is becoming more

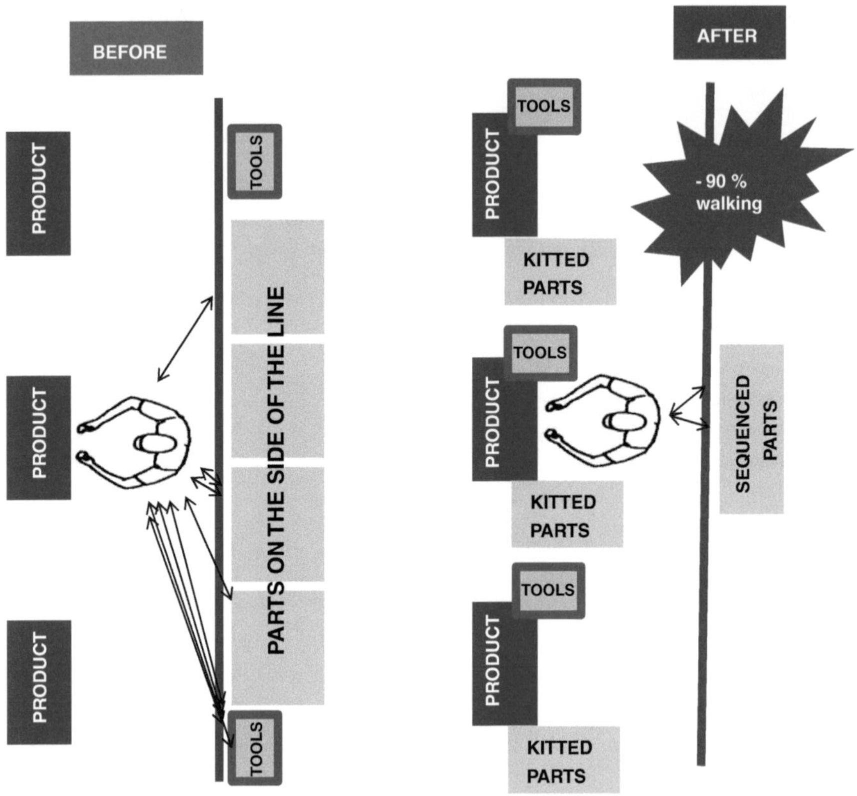

Fig. 4.3 Example of reduction of walking time on an assembling line

relevant day by day. For Western European productions, the impact of labour is, for the Final Producers, about 70 % of the final cost of production, including all the value added along the "supply chain" for components, equipment and tooling and services provided.

Once again, we remember how much the availability of labour is relevant to operative planning, so as to avoid misalignments between production capacity and product sales.

Finally, the labour factor is the "critical resource" for the competitiveness of productive systems for its influence on both « make and buy » costs and the capacity to answer the customer's demand.

For all these reasons, let us conclude this section by again pointing out the parameter generally used to measure and compare different plants, operating homogeneously in industrial activities and product range, in terms of **global labour productivity**, which means analysing the **individual annual labour productivity** (see beginning of this section).

This synthetic parameter depends mostly on the standard working times (ST) and consequently also on the automation level of processes. It also depends on labour efficiency (η) and the indirect labour incidence (i), but also on the number of working hours established by the contract (without considering holidays), as well as absenteeism for accidents and sick legally recognised permissions… In "benchmarking" research for manufacturing systems, it is important to use this factor when the different realities are comparable. To obtain an objective analysis, it is important to consider the data based on the product characteristics (dimension, complexity) and examination of the verticalization degrees of the production ("make-or-buy" configuration), correcting opportunely if necessary.

The above-mentioned productivity indicators, associated with unitary labour cost (payment and taxes), influences strategic decisions on new and existing sites for production activities, following the criteria discussed in Sect. 1.2.

To conclude this section, it is important to underline that labour productivity is not separate from product quality. In fact, a well-engineered process with efficient automatic systems contributes to optimization of working conditions, even from an ergonomic point of view, and at the same time improves product quality. So, productivity and quality of labour improve together.

4.5 Input Data for Operative Control

The Standard Productive Capacities of the technical systems n (SPC_n) and their Standard working Time ST_n are the basic input for Transformation Cost, both for the predictive phase (budget) and the accounting phase. The transformation cost and direct material trends are normally checked monthly, comparing the actual progressive accounting of the month with the budget forecast; this helps to direct corrective actions and makes sure the forecasts arc rcviewed every 3 months.

For direct material cost, the analyses are run according to the guidelines specified in Sect 2.5, in strict co-operation with the Product and Process Engineering Departments.

To check the transformation cost and accounting properly with product sold (cost accounting), according to "best practice" , it is possible to use two alternatives:

1^ MODE

The activities are grouped into *cost centres,* made by units manufacturing products that can be considered technically homogeneous. The transformation cost is accounted for single products based on productive hours of manpower used, with reference to the standard working time ST_n. Activity levels correspond to the standard working hours necessary to develop the production plan: PAV during the predictive phase (budget), DAV during the accounting phase.

The transformation costs and consequently the hourly cost (*labour burden*) are composed of three parts:

(a) The first part is the so-called **variable part,** because its annual amount is proportional to the activity volume (PAV, DAV). It includes the following factors of the cost per unit:

  (a1) direct labour working hours (ST_n/η), to which apply the average hourly cost for labour (payment and taxes);
  (a2) incidence on (a1) of the costs of direct material losses (rejects due to defects in the transformation process, over-usage in spite of standards in the bill of material);
  (a3) incidence on (a1) of the costs of consumables and tools, connected to the transformation process;
  (a4) incidence on (a1) of costs of energy and power supply in general, connected to the transformation process.

(b) The second part is the **fixed part,** because its annual amount is independent from the activity volume; it influences the *burden*, with reverse proportion to PAV (in the predictive phase) or to DAV (in the accounting phase). This incidence reaches the minimum point when the production plan fulfils the annual productive capacity and includes the following cost factors:

  (b1) depreciations related to the cost centre;
  (b2) fixed costs necessary for the functioning of the plant, even if charged to cost centre (overheads, technical and logistic departments, general equipment conduction, annual technical assistance, insurance…).

(c) The third part is the so-called **half-variable part,** because it has a "step" trend in respect to activity volume, in relation to the working shifts planned (PWT). Its incidence on the *burden* is minimal when the production plan fulfils the productive capacity of each weekly shift and includes the following cost factors:

  (c1) indirect labour, partially considered proportional to the activity volume PAV, DAV, partially considered fixed, considering what is specified in Sect. 4.2;
  (c2) maintenance costs, deployed in fixed and variable costs based on statistical analysis, according to criteria that will be specified later in Chap. 6;
  (c3) material handling costs, not included in the manufacturing cycle, deployed in fixed and variable costs based on statistical analysis, according to criteria that will be specified later in Chap. 5;
  (c4) energy and power supply costs (lighting, environmental air cooling and conditioning…), the amount of which depends on the number of working days and shifts.

Comparing the three above-mentioned cost factors (a, b, c) with the budget forecast and with the previous financial balance, it is first necessary to quantify the differences in the activity levels (DAV compared to PAV). This calculation derives from the quantity of product delivered to the final product warehouse (volumes and mix), and shipped out for the standard costs, to which it is necessary to add or subtract:

- distinctions due to technical differences in the product, introduced as per request by the Marketing Department and transferable to the profits;
- distinctions due to activity transference (to or from productive units).

Furthermore, for each one of the three cost factors (a, b, c) it is necessary to quantify the variations that have occurred or are foreseen as likely to occur in spite of the standard cost, due to the purchasing price of materials and services, the labour costs and the energy service fees. These variations are considered as " price effect" and include the effectiveness of negotiation.

Having separated variation due to, respectively, "volume-mix effect" and "price effect", the variation attached to each of the cost factors is determined by difference (efficiency or inefficiency in management).

For this purpose, it is necessary first to examine the standard working times ST_n, distinguishing the following variation causes:

(1) product/process technical changes necessary to improve/fix the quality level of the product (typically these changes worsen the cost);
(2) standard working time and investment reduction specifically approved for reducing the transformation cost (typically they are variable cost reductions balanced by higher depreciations);
(3) standard working time reduction for improvement of the process, not requiring specific relevant investments (they are variable cost reductions).

Then, it is necessary to examine the variations due to other cost factors such as: direct labour losses $(1 - \eta)$, incidence of indirect labour (i), direct material losses (deviation compared to the standard requirements in the bill of material), utilization of specific power supply and consumables, outsourced fees for production process services...

By analysing the nature (positive or negative), the make-up and the causes of these variations, both in the predictable phase (budget) and the final balancing phase, it is possible to understand the economic effectiveness of the production processes management.

2^ MODE

When the manufacturing systems are using hard automation solutions, the cost centres are organized by "homogeneous groups of equipment" or by "one-piece-flow integrated systems". The transformation cost is assigned to single products based on hours of utilization of the machines/equipment, referring to the specific standard productive capacities SPC_n and applying the relative hourly functioning cost (*machining burden*).Activity levels correspond to Net Planned Working Time NPWT.

As we have seen in Sect. 3.2, the standard productive capacity SPC_n is relative to each of the products *n in charge* and corresponds to what is obtainable if operating without breakdowns and managerial failures.

Not using the available productive capacity (1−U) determines the amount of "unabsorbed fixed costs". To evaluate the transformation cost trend, it is necessary to compare first the machine load (NPWT = net planned working time), foreseen in the budget, to that actually scheduled. The variation of cost due to the activity levels (utilization degree of the systems U) can be distinguished from that due to the overall equipment efficiency (OEE). These parameters are measured as described previously in Chap. 3.

The analysis of the utilization of cost factors, for the products actually in manufacturing, can also be achieved as described in the previous point, even if their incidences change, being activities with a higher impact on depreciations and maintenance costs.

Just as necessary as the building of the cost centre is that of the "manufacturing homogeneous systems", to which it is possible to apply the above-mentioned analysis criteria, although this method is more complex than the previous one. For this reason, it is not used in the Final Assembly Plants, where it is preferable to assume the standard working time ST_n as reference for the accounting of transformation costs to the products (1^ mode).

The 2^ mode is, however, predominantly used for economic control in the manufacturing of "capital intensive" components, which are technically homogeneous but with different part numbers and destined for more customers. In this way, the cost of machinery/equipment is properly assigned, clearly demonstrating the negative economic impact of the non-utilization of productive capacity. As a consequence, it is possible to decide if it is convenient or not to take other orders, even if with a minor profit margin in spite of normal rates, to make better use of the production capacity.

These criteria and methodologies for financial analysis fall under the module of "Finance for the Enterprise". The following diagram solely represents the typical trend of transformation unitary cost versus planned working time (corresponding to activity levels). In the given case study, the available working time is equal to three daily working shifts (22 h/day). This amount of working hours is typical of "capital intensive" automotive productions.

$$TUC = UVC + \frac{AFC}{(AU)} + \frac{SFC}{(AU)} \cdot \sum_{n=1}^{3} \frac{n \cdot SFC}{(SU)} \text{ where:}$$

AWT available working time (22 h)
n number of scheduled daily working shifts
AU degree of utilization of the available annual productive capacity
SU degree of utilization of the available productive capacity in each shift n
TUC transformation cost per unit
UVC variable cost per unit
AFC annual fixed cost incidence, with full utilization of productive capacity
SFC shift fixed cost incidence, with full utilization of productive capacity.

By analyzing the diagram in Fig. 4.4, it becomes clear how important it is to concentrate work in shifts, avoiding open, unfilled shifts as much as possible and using, if necessary, overtime.

Let us conclude this section by emphasizing that the so-called "continuous improvement" in labour and system productivity can be achieved only through the co-operation of the Production Management, Manufacturing Engineering, Maintenance and Logistics Departments.

The Finance Control Department is in charge of ensuring that "input" is complete and coherent, during both the budgeting and final balancing phases, using the necessary "auditing" activities and working up a constant monthly, quarterly and annual "report" on achieved results.

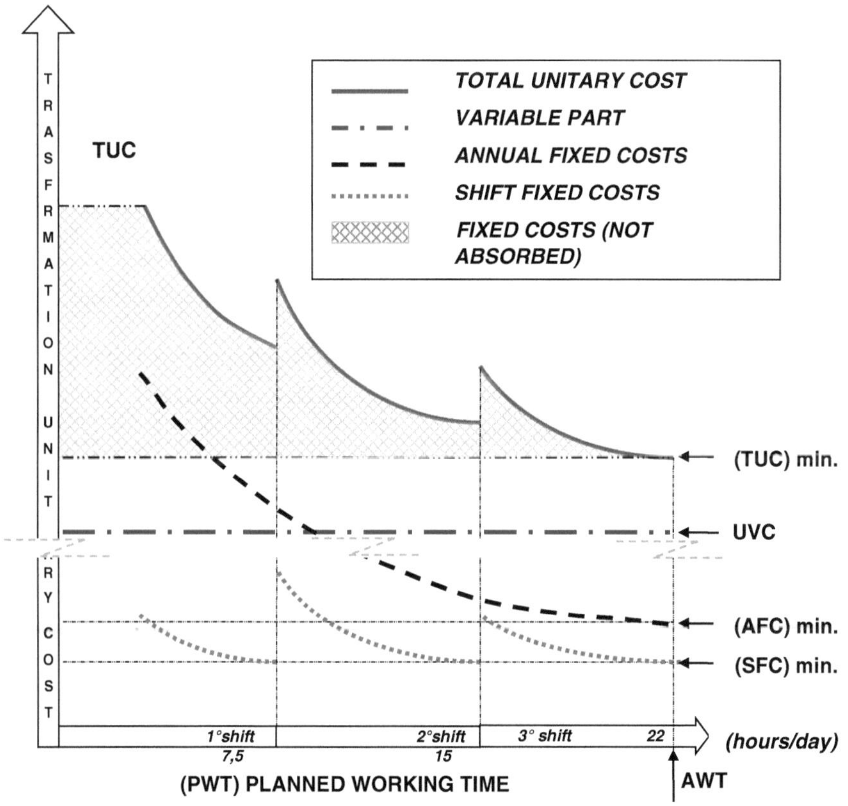

Fig. 4.4 Transformation Total Unitary Cost (TUC) trend in relation to Activity Level variations

4.6 Assignment of Tasks and Workload Balance

To better understand the assignment of tasks and workload balance, we have to analyse real case studies.

So, let us examine the criteria applied to a car's final and powertrain assembly, these processes being particularly "labour intensive" and very complex from a management profile. We will focus first on the issue of the assignment of tasks and workload distribution for a car's final assembly line and engine and transmission assembly lines. To do that, we will need the following technical information:

- elementary operation job descriptions, easily understood by the workers and possibly supported by a local terminal connected to the central information technology shop floor system;
- operation time required (standard working times ST_i) and priorities for the sequence to be followed, considering assembling needs and self-control;
- constraints in terms of workstations, due to component supply and tools (technological layout).

First of all, it is necessary to determine the required pace for the line to be able to comply with the productive operative plan, which corresponds to HVP (see also Sect. 3.2 for equipment productivity definition) :

$$\mathbf{HVP} = \mathbf{QA}/(\mathbf{AWT} \cdot \mathbf{A}), \quad \text{where :}$$

QA is the quantity of product assigned to the working shift, according to the production plan;

AWT is the available working time (hours per shift duration, without collective breaks);

A is the technical availability of the "system", according to the definition in Sect 3.5;

HVP is the hourly pace to be set on the line to satisfy the production plan, assuming all manpower required is available and that direct materials will be supplied regularly to the line, without any interruption

Based on the operative assembly sequence and the pre-determined standard working times, operations to be assigned to single workstations, interconnected with a one piece flow concept or separated by buffers, are determined, specifying tasks and the number of workers needed.

It is important to ensure that "finished operations" will be assigned at each workstation and all along the line, with each stage of the product being easily self-checked during assembly. In this way, each worker or group of workers can be responsible for quality as well as quantity. And, indeed, quantitative results (individual and through teamwork) can be checked automatically as well, using the actual information technology systems to detect and control (process traceability).

To optimize labour productivity and ensure the quality level, it is important to have a continuous and stable workload for each of the shifts. For each of the workstations or section of the line, it is necessary only to assign the number of workers strictly necessary for the operation, based on standard working times for the operation. The ideal workload can be achieved by complete realization of working capacities, which is when the following "balance point" is reached:

$$OT_i = NW_i \cdot WCT$$

where:

OT_i is the operative time, related to the workstation/section of line *i* (min); it is different from the standard working time actually assigned to the operators, because it does not consider physiological rest breaks, made by individual replacement or collective line stoppages between the two shifts; to obtain the operative time for every workstation, it is necessary first to deploy the standard time for all elementary operations required, and then aggregate them by workstation or line section in order to narrow as much as possible the gap between total time determined and a multiple of the Working Cycle Time that sets the pace of production or to a multiple of it being aware that:

$$ST_{tot} = \sum_{i=1}^{N} OT_i$$

where

N is the total number of elementary operations included in the considered process;

NW_i is the number of workers required simultaneously to operate at each workstation or section of the assembly line;

WCT 60/HVP is the working cycle time of the line (min).

In real situations, the priority constraints and time sequence of single operations do not allow for obtaining such an ideally balanced workload distribution for each one of the interconnected workstations or sections of the line.

To ensure that workload and working speed will be regular, especially for quality levels, we need to proceed in the following way:

1. *Workload balance*

   For the different planned activity levels (volume and product mix), activity should be divided, as much as possible, into the ideal workload for each workstation or line sector (*i*); this is called the "balance point".

   To do this, possible alternative solutions are analyzed (changes in operation sequence, combinations and rotations). Each solution must be naturally coherent with the manufacturing engineering plan and with ergonomic and layout constraints.

2. *Technical saturation evaluation*

Once the above analysis has been completed and the best solution defined, it is possible to evaluate the effectiveness of this activity by verifying the degree of technical saturation for the staff dedicated to production:

ts technical saturation degree = $\Sigma_n(\mathbf{OT}_n \times \mathbf{QA}_n)/(\mathbf{AWT} \times \mathbf{NW})$, where:

OT_n is the operative time for each product n (measured in standard working time, without physiological breaks);

QA_n is the quantity of product n planned in the working shift;

AWT is the daily available working time measured in working hours (without collective breaks);

NW is the quantity of workers assigned to the line (without substitutions needed for individual physiological breaks).

We emphasize that:

Ts 1 means that we have "ideal balance" for the workloads;

ts 0.9 means that the workload balance, considering the actual staff, admits an average activity level 10 % lower than the ideal

3. *Line staff determination*

Let us consider a real case in Italy.

As a consequence of union agreements drawn up in 1973, but recently re-negotiated according to improved working conditions and still abided by in some Italian plants, there are some constraints for the workload on assembly lines:

- for line speeds within 15–60 cycles/h, the average technical saturation degree (ats) allowed in the working shift is lower than 88 %;
- for line speeds higher than 60 cycles/h, the average technical saturation degree allowed is lower than 84 %.

  This agreement also establishes the parameters of permitted breaks, in particular for general assembly lines and, more generally, for all multi-station production systems directly connected. These breaks are forty minutes per shift and can be assigned individually, without stopping production, or collectively by stopping production.

  In the absence of specific agreements, it is still necessary to organize the workload, including physiological breaks, according to the work analysis criteria mentioned in Sect. 3.3.

  Considering collective physiological rest breaks (line stoppage) and individual breaks (worker replacement), the average technical saturation degree for the dedicated staff can be evaluated in the following way:

$$ats = ts \cdot \frac{NW}{NW + RW} \cdot \frac{PWT}{PWT + B},$$

where:

PWT is the planned working time in the working shift (without collective breaks)

NW is the total number of workers assigned to the line (without replacement workers)

B is the time for line stoppages due to collective breaks

RW is the number of workers assigned for replacement (to cover normal individual physiological breaks).

4. *CAPE analysis*

   For very complex assembly lines (cars, industrial and commercial vehicles final assembly and powertrain systems assembly), the major difficulties in balancing the workload are due to:

- sequence of operation constraints;
- layout constraints, especially due to specific equipment positioning;
- product mix variability and composition;
- workload range oscillation allowed in consequence of the spaces available for "shifting".

  The technical saturation degree for the working staff (ts) corresponds to the average value obtainable during the working shift. During the production program, some products (n) with a higher or lower operative time for the workstation or section of line can be introduced; in this case, temporary excess or lack of saturation, compared to the average value, can occur.

It is important to contain the "temporary over-saturations" within a 10 % limit, by temporarily shifting positions on the line and alternating longer operative phases with shorter ones.

When the assembly cycle is made of several interconnected phases and centred on many different products (more models and versions and many options), it is necessary to use CAPE methodologies, which allow for the simulation of other possible configurations, searching for the best level of technical saturation (ts).

In conclusion, the criteria for workload setting must provide a composition and sequence of product mix rightly balanced during the same working shift. This composition and sequence must correspond to the average standard working times measured by worker and workstation and should be coherent with the standard cycle time set according to the calculation mentioned earlier so as to reach the balancing point (ts = 1) as much as possible.

In assigning the working staff, it is also necessary to add the required RW of workers for the replacement of the workers NW that are operating on the line, for when they need individual physiological breaks (RW is generally a number within the 4 % of NW).

To summarize what has been described up to this point about the assignment of tasks and workload balance on an assembly line, the macro phases to be followed for maximal achievement on the job are:

1. pursue the ideal workload at every workstation (search for a OT_i at every station as a integer multiple of WCT)
2. obtain the real workload
3. check the technical saturation at every station
4. calculate the number of workers needed
5. balance the workload through possible adjustment between working stations (considering the sequence of the elementary operations to be respected and their length), by linking this activity to NVAA reduction as well
6. decide line break policy to be adopted and estimate average technical saturation

4.6.1 Considerations on Work Organization for Assembly Lines or Parallel Workstations

To improve working conditions and simplify the assignment of working tasks, in view of such restrictions on saturation, many *carmakers* have been trying to switch from an interconnected multi-station system to flexible modular systems. In particular, during the mid-'1970s (Volvo plants), technological layouts organized in "working islands", made up of parallel and "non-synchronic" workstations, were conceived.

The above-mentioned *isle systems,* unfortunately, were not ultimately successful for final assembly as a result of the tremendous complexity in logistics management they represented. The most successful experiences have been tested in the manual assembly of mechanical systems and main sub-systems (cockpit modules, air conditioning module, side doors, front modules...).

Let us try briefly to describe the main figures for the "working islands system":

- more operative phases are grouped together and more equal workstations are set in parallel to build a "homogeneous island" of activity;
- the technological layout is not linear, but divided in more "parallel sections", connected by programmable handling systems;
- within the same "system", more "homogeneous working islands" can exist, interconnected with each other in compliance with the assembly cycle and with the aid of automatic stations and buffers.

With such a technological layout, applying software programs capable of managing workloads and operative flows, some advantages can be achieved:

- workers can slightly modulate the working speed during the shift, autonomously using the permitted individual physiological breaks, according to standard working times and activity level assigned;
- during assembly operations, products are not moving, improving operations from an ergonomic point of view;
- repetition of movement and relative muscular fatigue are reduced;

- by changing the productive programs, the number of workstations used (stability of the assigned tasks) can be changed, while assigned operations remain the same;
- enlargement of the assigned tasks helps to make the workers feel responsible and reduces the psychological and physical fatigue typical of short operations.

Conversely, some disadvantages have to be taken into account, such as the extra space that is taken up by more complex handling systems. In fact, it is necessary to manage the main component supplying system by kitting them through special automatic carts.

In conclusion, in choosing between line assembly and homogeneous working island systems, it is important to evaluate thoroughly, from a managerial and economic point of view, the following advantages and disadvantages:

1. An assembly system with a synchronous one-piece-flow allows for the allocation of components very close to the workstations, avoiding double handling. By increasing the operative cycle time, the standard working times tend to reduce through the reduction of movement, up to a limit lower than the operative cycle length (minimum sustainable cycle time); under this limit, it is impossible to assign any "complete operation" to the worker. In this situation, it is necessary to introduce a second parallel section of the line, not necessarily operating with the same working cycle time.
2. An assembly system with a non-synchronous modular flow allows for the enlargement of the number of operative phases assigned to the workers, ensuring better self-control, up to the upper limit of the assigned task; above this limit, the workstations become much too complex, with too many components to be assembled and too many different tools to be used. Compared to a one-piece-flow system, this method of organization allows for obtaining a better degree of technical saturation and avoiding the necessity of having additionally required workers for replacement, which is an obstacle for the traceability of responsibility.

Applying work analysis methodologies, the best and most convenient solution has to be clear, also considering the product typology and the required productive volumes. For this purpose, it is important to trace the correlation between the operative cycle time and the real working time (operative time with the addition of reduced workforce and physiological breaks). As an example, Fig. 4.5 compares assembly line and homogeneous working islands solutions, in the case of the final assembly of a powertrain system. In conclusion, the analysis shows the most convenient areas in relation to the operative cycle time considered.

In the case study above, when the "island system" gives an advantage of 11 % on the necessary labour Working Time (WT), we have the payback point for the major investment needed compared to the "line system".

In conclusion, the choice to set an "island system", instead of a "line system", should take into account:

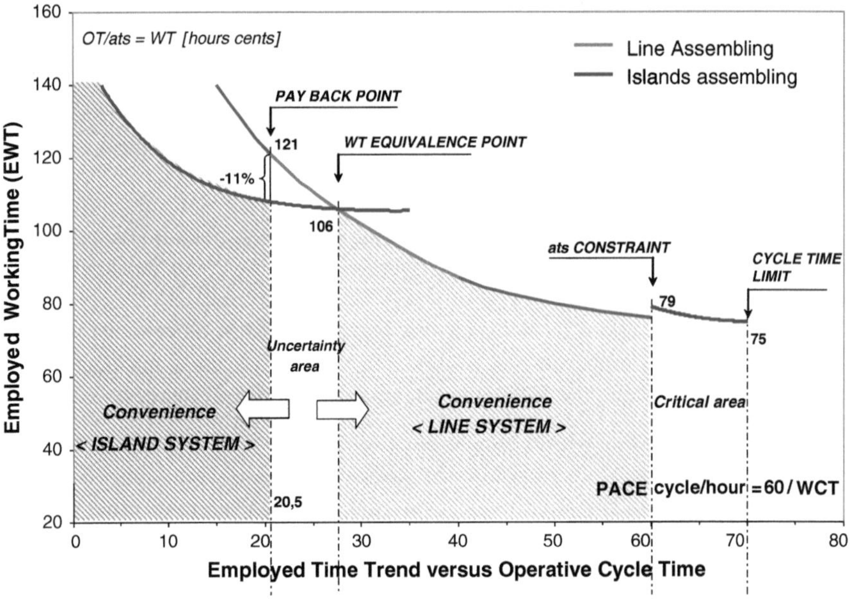

Fig. 4.5 Alternative technological solution comparison for powertrain system final assembly

- **attitude of workers towards performing "long tasks";**
- **normative and union agreements that enact constraints on the technical saturation degree (ts);**
- **space limits on the shop floors;**
- **payback point for the investment needed, thanks to better saturation of labour and the reduction of the defect rate in the final product.**

For the same production volume, the choices can be different, depending on the location of activities, the labour cost per unit and the attitude of workers.

4.7 Motivation and Rewarding of Employees

For the effective assignment of tasks and roles, it is necessary that:

- those in positions of responsibility have a deep understanding of their respective operations in the manufacturing cycle and the workload criteria as laid out in the previous section;
- technical documentation be easy to read, such as working schemes and procedures, assembly instructions and tools set, control plans…;
- necessary workers be assigned quickly, according to the requirement calculation, assuring good integration into the working team;

- workers be integrated gradually, with opportune training on the job and good support from the expert workers, with specific technical courses in advance if necessary.

For this purpose, Manufacturing and Human Resources Departments must cooperate closely.

Preliminary training, on-the-job training and the correct assignment of tasks and roles will be determinant in speeding up the "learning curve" (see Chap. 8); the speed of the start of production ramp-up, which contributes to the guarantee of the new products' "time to market" target, depends on it.

The progressive evolution in the organization of human resources, or so-called "lean organization", has determined and will determine in the future a deep change in the internal relationships and agreements with the unions. In modern enterprise, collaboration of employees is obtained by involving people in improvement activities and objective evaluation of individual performances, according to attitudes and on-the-job behaviour.

For production employees, it is important to evaluate:

- *availability and collaboration* in learning technical instructions and helping colleagues under training;
- *degree of attention* in avoiding errors affecting the quality level of the final product (self-control);
- *capacity to advance proposals* that contribute to the product and continuous improvement of the process;
- *punctuality and diligent continuity.*

This criterion of evaluation is simple and clear. Tellingly, it is the same by which the "artisan master" evaluates the apprentice!

Even in the biggest enterprises, these evaluations should be done by those directly responsible, supported methodologically by the Human Resources Department. In addition to the economic incentives, it is important to offer opportunity of professional growth to foster availability and collaboration.

Reprimands for bad behavior or poor work should be done with respect, even when necessarily severe. Those responsible must communicate directly and thus inject a sense of motivation into the disciplinary amends. To be effective, these amends should be "exceptions". If they become too common, suppositions for creating a positive climate for internal relationships will no longer exist. In this case, it is necessary to intervene from more than one direction:method and working conditions, internal communication, and involving people in target achievements and training activities.

If the "working teams" are unstable and there are too many of them, it becomes harder to evaluate, correct mistakes and foster training activities. For this reason as well, it is important to deploy production activities, avoiding critical situations.

Quality of working environment (spaces, lighting, climate…) helps collaboration and the achievement of quality targets. For this reason, modern enterprises engage in specific studies and investments.

Let us now talk of rewarding workers. Historically, the reward system was used for individuals "doing piecework": "the more you produce, the more you are rewarded". Again, this policy has been limited to small manufacturing enterprises and artisans.

In bigger manufacturing industries, where work is organized through "tayloristic" criteria, reward policies have been extended to larger groups of workers, interconnected and involved in the same productive process. The impact of these reward policies, based on labour efficiency (η), has been progressively reduced during the last decades. This fact is due not only to the opposition of the unions, but also to organizational changes that have occurred in the factories.

In fact, when resources required for the production plan are assigned and operations work in a "one-piece-flow", it is clear that the productive output—for the different stages of the process—is pre-determined and controlled on a standard basis. It is not allowed for a single worker to work at a lower speed than that required by the working cycle time (WCT), but neither that he work slightly faster, the latter for the purpose of evident regularity and stability of the production flow.

These considerations are relative to all mass production of complex products such as cars, commercial and industrial vehicles, motorbikes, agricultural means, domestic appliances….

Considering all of the above, it is important to consider that the reward policies of economic groups are very useful to the enterprise when they are based on indicators that quantify performances clearly and realistically. Referring to the previous definitions, some group rewarding factors can be:

(a) overall equipment efficiency (OEE);
(b) degree of productive labour utilization η / (1+i);
(c) quality indicators and cost of non-quality.

Another reward policy could be a collective annual prize, based on economic results within the same Business Unit or at the corporate level.

This last reward system—generally called a result prize—is applied by very big enterprises and normally is the object of specific union agreements, based on the logic of involving employees in corporate economic results.

The effect of these rewards on salary depends not only on the enterprise's policies, but also on agreements with the unions in accepting that a variable quote of the salary be linked to the enterprise's economic results (including the risk of withdrawing in case of failure).

Economic rewards are a part of the larger theme of participation and motivation of human resources to achieve a company's results. This argument is mainly a topic of Company Organization and Human Resources Development. Here, we only underline how to prevent the risks derived from an insufficient motivation of the employees due to repetitive and uninvolving tasks.

First of all, it must be considered that a single error due to lack of attention or sense of responsibility can generate a serious quality defect or production stoppage, or a delay in delivery. So, the risks for the company as a result of worker disaffection can be very serious. From this, we can derive the importance of obtaining collaboration from all employees, fostering a spirit of teamwork to help prevent critical events.

A good information technology system on the shop floor helps establish a sense of responsibility, involving employees in the presentation and evaluation of results achieved month by month, in terms of quantity and quality of the products and discussing customer problems (internal and external) due to delays in delivery or manufacturing failures.

Finally, we map out in seven points the best ways for a good leader to motivate his charges to engage in performing their tasks in the production process:

1. informing and training workers to perform their assigned tasks properly by applying the working procedures;
2. determining the right working tools and personal protection devices and initiating their use;
3. prompt attention in detecting errors and helping the workers to correct prevent them;
4. assuring that safety and sanitary regulations be respected and preventing risk of accidents and damage to workers' health;
5. establishing consistent attendance and behaviour on the job, adopting disciplinary amends whenever regulations are repeatedly disrespected;
6. presenting and evaluating results obtained by the working team (quality level, efficiency…);
7. collecting proposals for improvement and responding to them when possible.

To do this properly, managers and those responsible for production should have an attitude towards their relationship with the various levels of workers and overseers that is both sensitive and severe as the situation demands, but always based on fairness of behaviour.

The paradigm of a "lean organization" includes the concept of a "lean hierarchical structure", to delegate more responsibilities and cultivate a more competent approach to solving problems in the organization, avoiding the need to address them higher up in the hierarchy.

For this purpose, managers and those responsible should be "teamwork" oriented, with constant care towards the human resources engaged in company processes and demonstrating "leadership" based on employee involvement and constant determination in target achievement.

Chapter 5
Manufacturing System Management and Maintenance Criteria

5.1 Plant Manufacturing System

By "Plant Manufacturing System", we mean all the equipment and working means necessary for production. Machine, equipment and tool typologies are different from process to process, as mentioned in Sect. 1.5.

In general, for manufacturing industries, Plant Technical Systems are the main "capital assets". Accurate technical management of these systems is fundamental.

The information technology support used for planning and controlling investment initiatives is based, according to "best practice", on the PDM logic described in Sect. 2.3, obviously referring to the components of the Plant Technical System and not to the product parts. Each system's project corresponds to an analytic structure of the activities (WBS = Work Break-Down Structure), correlated to each "module" included in the technological layout and referring to the single operations of the working cycle.

The above-mentioned WBS also includes the "responsibility matrix", which defines the competencies for developing the single items according to the procedures applied to project initiatives, distinguishing internal from external activities, the last being charged to external suppliers of equipment and tools.

Based on executive layout, investment items are defined and classified, assigning the proper values derived from the single project initiatives.

The file is a "working means database", including the following information necessary for controlling investments and assigning depreciations to the single cost centres:

(a) plant/unit/cost centre to which the working mean is assigned;
(b) serial alphanumeric code, including information on the type of mean and the progressive serial number necessary for the identification tag;
(c) initiative's code or investment project, the purpose for which the investment has been delivered;
(d) manufacturing cycle for which the mean is used, whenever it is specific to the productive process, otherwise the auxiliary function for which it is used;
(e) date of assignment to the productive unit and introduction into the database;

M. Gobetto, *Operations Management in Automotive Industries*,
Springer Series in Advanced Manufacturing, DOI: 10.1007/978-94-007-7593-0_5,

(f) origin value and depreciation plan, according to the investment initiative;
(g) remaining value, without the yearly cumulated progressive depreciation (net book value);
(h) possible re-evaluation due to extraordinary interventions and relative mandate depreciation plan.

This database also includes the working means given for loan to external suppliers, but which is still the property of the enterprise itself.

To define the economic depreciation plan, the plant technical systems are formally divided into the following functional macro-categories:

(1) **Product Dedicated Equipment**
Special Machine Tools and Product Dedicated Equipment for specific product design production and not economically transformable. The related depreciation plan is generally for the short-medium term, being linked to the industrial life cycle of the single products.
(2) **Standard Process Equipment**
Machine Tools and Standard Process Equipment, usable for a large scale of products and economically transformable for new models, anticipated in the product range plan.
(3) **Plant Services Equipment**
Machine Tools for Equipment Construction and for Working Means Maintenance, Complementary Equipment for plant services, Information Technologies Systems, Measuring Machines, etc.
For "plant technical systems" in points 2 and 3, the depreciation plan is normally determined in the medium–long term, within 6–10 years, according to the following predictions:

- technical life, without extraordinary maintenance intervention;
- probability of technical obsolescence, due to the development of more convenient technological solutions.

(4) **Plant Facilities**
These also include buildings and all the facilities; the depreciation plan is for the long-term (more than 10 years, according to the administrative rules for the enterprise's balance).

The above-mentioned economic depreciation criteria stretch to avoid "residual values" of the original investment, when the items will no longer be used. They correspond to a prudential logic in drawing up the final economic balances.

For each of the items in the working means database, according to the responsibility matrix defined by the Steering Committee, the Managers must assure the following:

- verification that the working mean is complete, functional and safe for use when finally delivered from the supplier and put into production (according to technical specifications);
- inclusion of each item in the working means database;

- verification of the availability of warranty from the supplier side;
- guarantee of the correct usage, normal trustworthy maintenance and traceability of the working mean up until the time they are put in charge of the productive unit;
- advisement of when the working means is out of production, to begin procedure of transfer or disinvestment of the item.

In case of disinvestment of an item that still has a residual value, it is saved as a negative value on the final balance. In case of transfer, with a higher incoming than the residual value, it is saved as a positive value on the final balance. Periodically, the Finance Administration Department will verify the situation in collaboration with Technical Managers to assure the coherence of financial data with physical data.

5.2 Reliability and Maintainability of Equipment

Before starting with the concept of maintenance, it is important to define what we mean by machines' and equipment's ***reliability degree*** and ***characteristics of maintainability***.

A certain machine or manufacturing system can be considered fully reliable when it is able to perform its own function consistently and in a manner coherent with its established use during the requisite time for a given project.

From a statistical point of view, the reliability degree is the probability that it will work correctly and without failure during the progressive utilization time (t):

$\mathbf{r(t) = 1 - f(t)}$, f(t) being the probability of failure

The reliability degree depends on the following factors:

- **Intrinsic Functional Reliability**, determined by project quality and robustness, components and system quality in assembly. Techniques used for this purpose are: *robust design, FMEA analysis, transformation process simulations (during design phase), component and system functional testing techniques (during construction and try out phase).*
- **Duration Reliability**, linked to usage of mechanical and mechatronic parts and fatigue factors. Techniques used to establish and extend the life of components are essentially based on the right choice of material and calculation tests, through CAD-CAE methodologies and the application of standard solutions, tested with positive results.
- **Utilization and Maintenance Reliability**, determined by correct usage of the "systems" and effectiveness of maintenance processes. It is important to consider that the productive operations lead to a risk of failure; in the following sections, we will approach maintenance methodologies that help to improve this factor of reliability.

During the start-up phase, up to the final working speed of a system, the more influential reliability factor is the intrinsic one, while the duration reliability factor is strongly influential on the maintenance cost, if the production process is set, and determines the technical life cycle of machines and equipment.

The total reliability degree is the product of the three factors.

Let us now introduce the concept of **equipment and machine maintainability**: it consists of the easy inspection and substitution of overused parts and the correction of possible failures with simple and reliable operations.

The maintainability degree depends first on the design and development setting of the project and on the application of standard solutions, tested ahead of time in a collaboration between Constructor and User. To establish premium levels of technical reliability and availability (**A**), it is important to evaluate the possible failure modes, the associated probabilities, and the subsequent effects (FMEA methodology) during the project phase. After this analysis, the right "project reviews" are set, making construction solutions more robust and inspecting activities on critical components easier, assuring quick intervention for the substitution of damaged or used parts during the normal technical life of the equipment.

So far, in logical order, the main criteria that an engineer must consider for the design of easily maintained machines are:

1. Improve **detectability** of breakdowns by designing easy man–machine interfaces able to give maximum details about how a particular failure occurred (type, module/component position, instruction on how to access the damaged part…)
2. Improve **accessibility** to the machine: once the position of the affected module/component has been identified, access must be easy for the maintenance operator; this is possible thanks to two main approaches in design:
   - layout optimization (locate critical module/component in easily accessible areas)
   - facilitate the tear down of safety repairs by creating fast, user-friendly solutions for dismounting and mounting
3. Design by **modularity**: machines must be designed such that maintenance crews can operate solely on the relevant modules during a failure crisis thus speeding up maintenance operations; once the dysfunctional module has been replaced, it can be removed, analysed and subsequently reconditioned to work again through repair of the damaged component, all done remotely
4. Design by **standardization**: usage of standard components can improve the maintainability of the machine by:

   (a) facilitating the training of the maintenance crew and creating conditions for a quick response (interchangeability of skill trades)

   (b) facilitating the management of spare parts (reduction of supply lead times and inventories).

Even for equipment and machines already under production, it is possible to improve maintainability by acting as follows:

- improving the interface between man and machine through monitored instructions for driven intervention and time-based maintenance;
- out of machine "pre-setting" techniques applied for tool changes, to shorten set-up-time;
- improving the removal and re-positioning operations on safety barriers and repairs;
- partner tool adoption to the quick removal and re-positioning of bulky parts under failure, by operating in safety conditions.

For each module of a "Plant Technical System", the failure modes can be different, random and periodic. The techniques for cause and effect analysis—specifically discussed in Chap. 8—allow for selecting the critical phenomena and adopting specific countermeasures. This method is particularly important during the start-up phase: the root causes of the failures are verified, and eventual deficiencies in mechanical and electronic control systems are fixed (with particular attention to sensors and automatic system control software). In this way, certain types of failure and their frequency are minimized. In between, the intervention for corrective action is improved, reducing the length of technical breakdowns (maintenance learning curve).

Once the start-up phase is completed, the frequency index F_n, considered physiological for a certain typology of failure n, is estimated on a statistical basis. These analyses must be performed on a significant and extended temporal scale, such as: working week or month, or a certain progressive working time used for production (ex.: 100 h).

F_n = 100 ·/$MTBF_n$ is the average frequency for breakdown n referring to 100 h of work, where **$MTBF_n$ (mean time between failures)** are the average working hours between the failure events n, without considering inactive time for other causes.

The statistical data **F_n** is significant for the probability of breakdown and depends on the total reliability of the system.

The second important group of statistical data is the acronym **$MTTR_n$ (mean time to repair and restart)** and corresponds to the average downtime for a machine, expressed in hours, necessary for repairing a breakdown (n) and restarting production. MTTR includes:

- time necessary to detect the breakdown and maintenance intervention management;
- time necessary for repair with machine stopped;
- time necessary to restart production, after necessary test for safety and process quality.

The **$MTTR_n$** can also be determined by statistical observation and depends on the "degree of maintainability" and "maintenance efficiency" (intervention speed). To perform correct statistical data collection for $MTTR_n$, abnormal waiting time, due to a lack of manpower or spare parts (management problems), should be excluded.

Frequency and length of breakdowns negatively influence the available production capacity **(APC)**, according to what was demonstrated in Sect. 3.5. The **percentage breakdown severity index**, referring to the observed working time (t), derives from the following calculation:

$$s(t) = \sum_{n=1}^{K} F_n \cdot (MTTR)_n / t \text{ where:}$$

t	is the progressive working time during the period, expressed in hours
n	are the several types of breakdown observed
k	is the total number of breakdowns observed
$MTTR_n$	is the "mean time to repair and restart" for each type of breakdown (n), expressed in hours
F_n	is the average frequency for each type of breakdown (n).

The above statistical observation allows for the selection of priorities in countermeasure and improvement activities. For this purpose, a pareto diagram gives evidence of the percentage impact of the several types of failure n, both in terms of frequency and severity. This methodology will be detailed in Sect. 8.5.

Referring to the working time diagram presented in Sect. 3.5, we remember that the technical reliability degree is given by the ratio:

$$\mathbf{R} = \text{PUT}/(\text{PUT} + \text{BUT}).$$

For a specific manufacturing system or stand-alone machine, the trend for **s(t),** relative to the progressive working time cumulated, is typically shaped on a "bathtub" curve, as shown in Fig. 5.1. On the y axle, we find the incidence of technical stoppage due to breakdowns, referring to the working time used for production.

As shown in the diagram, the *normal technical life* period shows a constant breakdown index during the given time if an effective maintenance process is assured, or slightly decreased if continuous improvement activity on the production

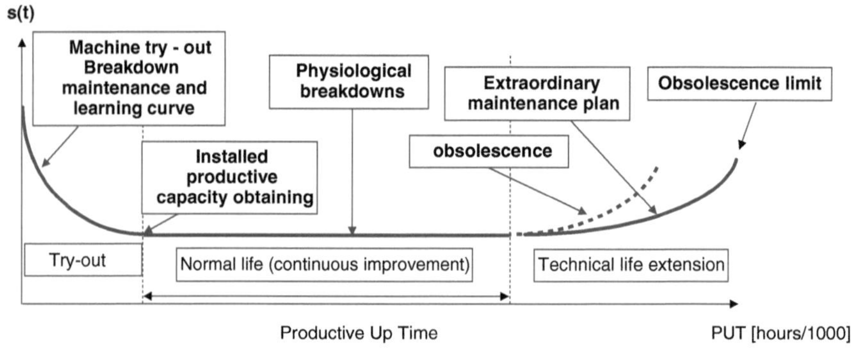

Fig. 5.1 Equipment's breakdown severity index in relation to technical life

system is applied, according to the logic that will be detailed in Chap. 8. During this period, breakdowns are caused by errors in utilization and by some physiological failures.

It is important to focus on the initial and final phases of the technical life of the system:

- ***Try-out phase***, characterized by several types of failure and high frequencies that decrease exponentially during the given time (electronic and mechatronic component with infancy illness); for this reason, the adoption of breakdown maintenance in this phase is suggested, in order to search for the right countermeasures by applying cause and effect analysis, which we will be dealing with in Chap. 8.
- ***Obsolescence phase***, during which breakdowns become progressively more severe as a consequence of structural components reaching the fatigue limit, with an exponentially increasing trend; this phase can be delayed by using intensive professional maintenance plans (predictive maintenance and extraordinary maintenance activities, with the substitution of critical structural parts…).
- ***Obsolescence limit***, which occurs when the system is no longer capable of guaranteeing the necessary quality level of the final products and/or the continuity of the production process; in many cases, technical obsolescence occurs for technological reasons, with the availability of more modern solutions making the employment of new systems more convenient, or as the result of the advent of more severe legal requirements or, finally, for technological needs of new products.

Before concluding this topic, let us examine the interference effect of technical breakdowns in the so-called "integrated production systems".

Having the same reliability in the single modules of operation in a system, that system's total reliability will be influenced by layout solutions and by interconnections adopted within the single stages of the system. Here, we approach the matter in a simple way, without using complex numerical models that should be based on the availability of statistical data, always hard to collect in a complex manufacturing system.

When the single modules are in **parallel**, with the possibility of working autonomously, even if in a unique integrated system, the total reliability degree for the system corresponds to the average value of the single reliability of each module:

$$\mathbf{R_t = (R_1 + R_2 + R_3 + \cdots + R_k)/k}$$

where:

R_t is the reliability of the system
R_k is the reliability of the module k
K is the number of modules in parallel.

Conversely, if the layout of the system is set on modules that operate in **sequence**, interconnected one to the other in a synchronous way, the total reliability degree is equal to the product of the reliability degree of the single modules:

$$\mathbf{R_t = R_1 \cdot R_2 \cdot R_3 \cdots Rt_k.}$$

Practically, this method of determining the reliability degree is valid if the breakdown distribution can be considered random and if, in case of concurrent events on two or more modules, there is no simultaneous repair intervention. In this situation, as it normally occurs, the failure of a single module causes the breakdown of the whole system; in other words, there is full interference of the technical failures of the single modules for operative continuity of the integrated system.

The above-mentioned interference can be reduced when the modules are interconnected by "flexible material handling systems", having appropriate dynamic inter-operational buffers. Even if this solution is recommended for increasing the total reliability degree, it is counterproductive for the continuous one-piece-flow concept that we will discuss in Chap. 6.

To calculate the optimal size of the "buffers", some considerations have to be taken:

(1) MTBF and MTTR for each module of the system should be known;
(2) additional investment needed for construction of the buffers, for all the alternative solutions considered;
(3) the obtainable gain in productivity level of the system by introducing the buffers (higher available productive capacity APC, higher overall equipment efficiency OEE).

The modern CAPE supports allow for the simulation of continuity conditions of productive flows, when the programmed technical stoppages are needed for set-up operations **(pu)** and when probable failure events *(n)* can occur, considering the frequency index F_n and the duration $MTTR_n$. In this way, during the project setting phase, alternative solutions for layout composition can be compared (modularity level and deployment of operations), searching for the best solutions for module positioning and for dimension of the buffers. These CAPE systems are also useful for defining the convenient situation of simultaneous tool change with the machine stopped.

5.3 Maintenance Management

The goal of maintenance is to maintain the original conditions of operation and safety necessary to fulfil production plans during the entire life cycle of a plant's technical systems.

Referring to the definitions given in Chap. 3, the technical parameters used to evaluate effective maintenance are:

- *technical availability of machines and equipment **(A)***;
- *stability of process and quality of final product **(process capability)***;
- *safety of work in terms of utilization and environmental emission, estimated according to **legal requirements**.*

From an economic-managerial point of view, maintenance activities can be divided into:

1. **Ordinary Maintenance**
   This includes necessary intervention for maintaining "plant technical systems" at normal functioning levels, according to quality, safety and productivity standards set by the project and consolidated during production.
2. **Extraordinary Maintenance**
   This includes all exceptional intervention for mechanical and structural parts of machines and equipment, necessary, after a long period of utilization, to avoid critical working conditions or to re-set equipment standards for modern technologies, consequent to new technical standards; with these interventions, the normal technical life of the equipment is increased as well as its remaining value (original investment without depreciations already accounted for).

Costs related to ordinary maintenance are accounted to the single unit or department in charge of equipment management and are included in the "standard transformation cost" of products. They are considered in the annual budget exercise and are updated monthly and yearly.

Conversely, investments necessary for extraordinary maintenance operations must be preliminarily evaluated from a cost-benefit point of view and then approved on a specific project initiative, to be depreciated in two or more yearly finance exercises, in relation to the lengthened life of the equipment involved.

The chart in Fig. 5.2 summarizes the main figures for Ordinary Maintenance Management for complex technical systems.

5.3.1 Autonomous Maintenance Activities (Simple and Recurrent Interventions: Type A)

This includes the interventions that can be performed directly by machine and equipment operators. These activities are simple and recurrent and must be done following specific standard maintenance procedures defined by Technical Departments (Manufacturing Engineering, Maintenance Engineering, Equipment Supplier) with the cooperation of operators. Typical examples of such activities are:

- process parameter control and adjustment;
- used tool and part calibration and substitution, when easily changed;

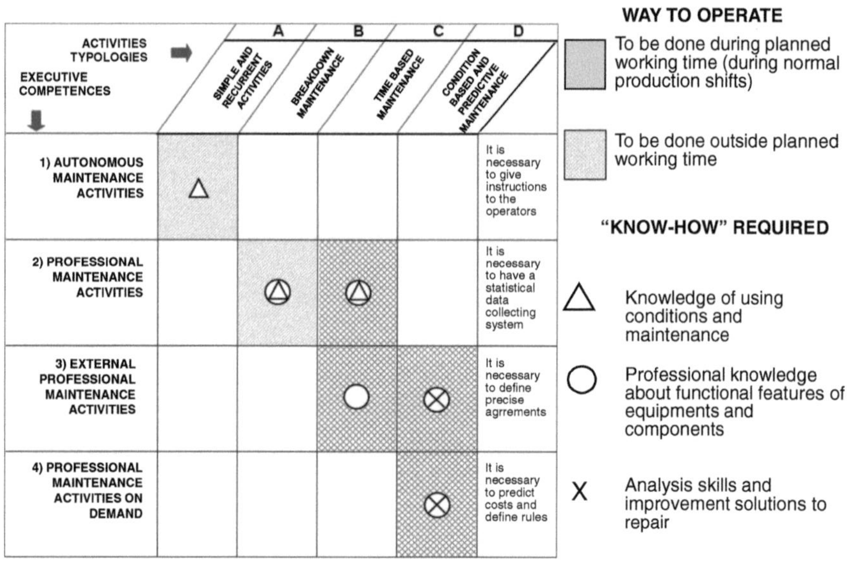

Fig. 5.2 Ordinary maintenance activities classification

- calibration and substitution of equipment components easy to access in safe conditions, for which any specific knowledge is required;
- changeover or intervention during the change of a working shift;
- lubricant filling;
- technical cleaning during the working shift.

Advantages of applying autonomous maintenance are:

(1) prevention of breakdowns due to the lack of basic conditions, operators being able to recognize weak signals from the behaviour of the working means;
(2) immediate correction of micro-stoppages, avoiding unnecessary waiting times for professional maintenance operators;
(3) containment of indirect manpower requirements for maintenance activities.

With specific reference to point 3 above it is important to notice that, since the utilization of direct labour for indirect labour activities like maintenance is costing, it is extremely important to determine which professional maintenance activities are convenient to be transferred to autonomous maintenance also from an economic point of view, by balancing costs sustained with benefit achieved.

To develop an effective autonomous maintenance system, it is necessary to train production operators specifically, establishing solid involvement and collaboration between maintenance staff and production workers, through sharing of the same targets (overall equipment efficiency ***OEE,*** functioning cost reduction, reject reduction…).

5.3.2 Professional Maintenance Activites (Type B, C and D)

These are charged to Staff Technical Departments, which should be proficient in:

- Analysing the root causes of breakdowns and determining countermeasures to effect in agreement with equipment constructor.
- Studying methods for intervention and setting necessary resources (personnel, working means, spare parts).
- Organizing resources locally to assure a fast response to requests for maintenance intervention.

Professional Maintenance can be directly managed by final producer plants, organizing a specific maintenance department, or renting the service from an external specialized provider, operating with a specific open agreement. Criteria for the selection of "make-or-buy" services are the same as those explained in Sect. 1.2. From an organizational point of view, professional maintenance activities can be divided into the following categories:

Breakdown Maintenance, on user's request.

For this type of intervention, it is necessary to have local specialist teams, close to the production units. The size of these dedicated teams is established by considering the frequency of requests on a statistical basis and intervention time required (MTTR), according to technical availability (**A**) targets for each of the planned working shifts.

It is important to know that "quick intervention" does not always solve the problem definitively; in this case, it is necessary to set a temporary solution, followed by a final definitive countermeasure, to be planned opportunely.

Time-Based Maintenance (TBM), to be performed as follows:

- ***in calendar terms***, possibly outside of the working shift so as not to interfere with production activity, when dealing with consumable materials (lubricant, filters, galvanic fluids…) that decline by themselves, or with equipment under legal safety prescriptions (lifters, high pressure tanks, fluid depuration equipment…);
- ***in working time terms***, calculated based on the progressive working time cumulated; typical such interventions are: substitution of used tools, machine parts under mechanical stress (gears and bearings, transmission chains and belts, mechatronic devices…). Additionally, these interventions should be done outside of planned working time whenever possible.

Maintenance standard procedure and frequency are defined based on statistical analysis and economic estimation. Data are more reliable when culled from standard solutions widely tested in the field.

We can observe that fatigue limits for machine parts under stress depend on the behaviour and dimensions of the forming materials. Eventual critical conditions, due to accidental failures or lack of maintenance, influence the frequency of maintenance intervention. Discounting these anomalous events, normal component life can be represented by a curve comparable to a gauss histogram distribution as represented in the following diagram.

The bigger the dispersion relative to breakdown frequency, the more severe the costs for preventive substitution of parts subject to fatigue stress. To choose the right frequency of intervention, it is necessary to search the convenience point on a statistical base, comparing costs for preventive maintenance to benefits derived from well-functioning equipment (fewer production losses).

Considering the distribution in Fig. 5.3, we can distinguish the following alternative intervention strategies:

1. Period of intervention WT/1, preventing 97 % of probable failure events; in the case shown, this leads to double the specific maintenance cost, compared to a breakdown maintenance strategy.
2. Period of intervention WT/2, preventing 85 % of probable failure events; in the case shown, this leads to an increase of 25 % of the specific maintenance cost, compared to a breakdown maintenance strategy.

It is more convenient to adopt strategy (1), oriented to a higher technical reliability, when the type of breakdown can affect final product quality or imply severe interruption of production activity, compromising the final level of customer service and leading consequently to high costs for functioning (final product stocks).

It is more convenient to adopt strategy (2), oriented to optimize working costs, when the type of breakdown affects the technical availability (**A**) without implying the risks of the previous point.

Condition Based Maintenance (CBM) and Predictive Maintenance, in consequence of specific controls on the working conditions of the equipment and on "process capability".

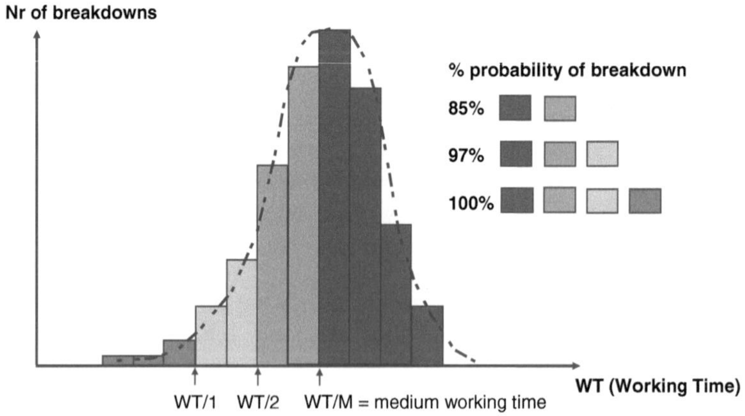

Fig. 5.3 Breakdown phenomena statistical distribution in relation to progressive working time

For this purpose, it is first defined as the necessary cycle of inspections with the relative frequencies, to be performed during the planned working time, in collaboration with the users. These inspections do not imply the disassembly of the machine's mechanical parts and are possible thanks to techniques based on process parameters controlled through specific "detecting tools" that work on equipment working conditions.

As a result of these inspections, technicians evaluate the necessity of predictive maintenance intervention, to prevent failures.

According to the best practices, the techniques most used for diagnosis are:

- working cycle analyser, interfaced with machine control panels;
- vibration analysis, thermography, infrared analysis, X-ray analysis, fluidic tests…;
- telemetric tests for geometrical trim of kinematic parts influencing "process capability";
- remote diagnosis systems, that allow for the use of Technical Assistance Centres, specialized for addressing the re-setting of interventions remotely.

The above-mentioned techniques should be opportunely integrated by statistical analysis in the field, as previously recommended.

The development of planned maintenance, under predictive inspection, can lead to consistent economic and technical advantages, but requires highly qualified employees and specific techniques, of the types mentioned above. For this reason, "carmakers" and Main Component Producers tend to delegate this type of activity to Equipment Suppliers themselves or to Specialized Maintenance Service Companies, through specific technical assistance purchasing agreements.

Let us conclude the argument in favour of maintenance activities organization, underlining the importance of studying "methods of intervention".

The starting point is statistical data MTBF, MTTR and the estimation of specialized manpower requirements for performing maintenance activities, having established a standard time of intervention. For parts with normal usage, frequency and length of intervention are defined in a predictive way, considering conditions for substitution and intervention cost compared to the benefits obtainable through advantages in the production process.

Relative to more complex planned maintenance activities, it is necessary first to define the methods, time and frequency of intervention required with accuracy, through specific statistical analysis, whereupon personnel requirement can be determined applying the criteria for working time and work analysis demonstrated in Sects. 3.3 and 4.1.

We need to emphasize that, for maintenance activities, it is impossible to reach the full saturation of dedicated labour, due to the fact that workload variations are relevant and long waiting times can be generated through lengthy analysis of breakdowns and the time it takes for spare parts to be supplied. To optimize the maintenance workforce, the saturation degree can be determined statistically, with a proper system for recording interventions. Generally, a 70 % degree of saturation can be considered acceptable, because it avoids delays for emergency calls and contains maintenance costs.

To reduce MTTR, we must act on the following factors:

- **increasing technicians' professional knowledge and improving operators' skills through training, on-the-job in the case of the latter;**
- **supplying the right tools for detecting failures' mode and for speeding up maintenance intervention;**
- **quick availability of spare parts.**

In relation to the last point, the logistic organization of the supply system is particularly strategic. Spare parts can be classified into the following categories:

(a) **Spare parts necessary for recurrent "back-up"**, which should be available at machine side to speed-up interventions; these parts are normally recoverable and repairable for further use as "back-up" parts.
(b) **Spare parts necessary for non-recurrent interventions**, the storage space of which should be nearby the equipment and machines on the shop floor and managed with a "minimum stock level" criteria, considering utilization frequencies and supply lead times.
(c) **Standard parts widely employed**, the storage space of which should be organized on two levels: the first peripheral level should be small and located near the equipment and machines on the shop floor, while the second level should be centrally located and able to serve more plants, operating to replace the parts withdrawn from the peripheral warehouse on the first level, normally at a weekly rate. In this way it is possible to assure a good service level and contain the relative invested working capital.

To simplify the process of supplying spare parts and to minimize working capital, technical solutions and automatic components must be standardized as much as possible (the same type of machine centre, robot and controlling devices, material handling equipment, etc., should all be used in the same plant).

Modern ICT allow for the automatic checking of stock levels in the warehouses in real time and for the organization of and supply plans for spare parts based on "minimum safety stock" criteria.

To purchase spare parts, it is normal to operate on open agreements, engaging the suppliers in obtaining short delivery lead time and then relying on them to be in charge of supplying the necessary stocks.

5.4 Correlation Between Cost and Maintenance Effectiveness

As seen in Sect. 4.5, maintenance expenses for manufacturing systems significantly influence hourly transformation costs (burden). This incidence is particularly high for metal transformation processes (foundries, steel metal printing, machining), reinforced polymer transformation processes and body welding and painting.

To plan and control these costs, it is first important to distinguish the part considered "variable" in relation to the activity levels (PWT and PAV) from the part considered "fixed" on a yearly basis, depending on the structural asset of the plant's technical systems.

On first approximation and in proportion to activity level (PWT and PAV), we can consider the expenses as variable for:

- consumable materials and spare parts used in maintenance, evaluated at the supplied cost;
- fees for time-based maintenance activities planned on cumulated working time and evaluated at contract price (if performed by external providers) or at hourly labour cost of Maintenance Department.

We can consider expenses linked to fixing maintenance garrisons set for assistance to production and emergency calls for each of the planned working shifts as "variable by step", in relation to the planned working time (PWT).

Conversely, we can consider expenses as fixed per year for:

- salaries for managers and technicians dedicated to maintenance;
- other technical and logistic services provided by external suppliers on specific initiatives;
- time-based maintenance activities on calendar normally provided by external suppliers, set at contract price.

In this way, it is easy to plan and control maintenance cost trends, analysing variation causes month by month and year by year, distinguishing:

- variation due to activity volume effect (PAV and PWT);
- variation due to purchasing prices and labour unitary cost;
- variation due to efficiencies.

The evaluation of maintenance cost trends should be done together with the evaluation of technical availability (**A**) and overall equipment efficiency (**OEE**) for the plant's technical systems.

We have to emphasize that, in searching for the best maintenance asset (organizational structure and methodologies for intervention), the following points should be considered:

- incidence of maintenance costs referring to total transformation costs, normally higher when accompanied by a quick response to emergency calls and a high level of failure prevention, with positive effect on technical availability (A);
- incidence of costs due to productivity losses, which depends on statistical data (1-OEE), according to what was laid out in Sect. 3.5;
- economic advantage obtainable from increase in equipment productivity, as a consequence of a higher available productive capacity (possibility of reducing working shifts, or increasing the sales plan).

To conclude this section, we will focus on the labour factor of maintenance costs, which we classified as "indirect labour" in Chap. 4; looking at all maintenance parameters, it is possible to estimate the maintenance labour force required for maintaining a production department by measuring the following data:

- frequency and severity of breakdowns by department
- frequency and duration of preventive activities recorded in the maintenance calendar (TBM and CBM activities).

Referring to the calculation of the Direct Labour Requirement (DLR) seen in Sect. 4.2, the calculation of Maintenance Labour Requirement could be carried out by the following calculations in.

5.4.1 Breakdown Maintenance Activity Volume

$$BMAV = \sum_{i=1}^{N} \frac{1}{60} \mathrm{MTTR}_i \cdot \mathrm{B}_i$$

where:

- MTTR_i is the Mean Time To Repair at the production department level measured in minutes;
- B_i is the total number of breakdowns occurring during the working period considered in the department for the typology of failure I;
- N is the number of different typologies of failure occurring in the department.

5.4.2 Preventive Professional Maintenance Activity Volume

$$PMAV = \sum_{j=1}^{J} \frac{1}{60} \mathrm{TBM}_j + \sum_{k=1}^{K} \frac{1}{60} \mathrm{CBM}_k$$

where:

- TBM_j is the time required for the Time Based Maintenance activities in the department recorded on the professional maintenance calendar, measured in minutes;
- J is the number of different Time Based Maintenance activities managed in the production department;
- CBM_k is the time required to execute the Condition Based Maintenance activities in the department recorded on the professional maintenance calendar, measured in minutes;
- K is the number of different Condition Based Maintenance activities managed in the production department.

5.4.3 Individual Activity Achievable

IAA = $\eta \cdot$ **IWH** $\cdot$ **(1−a)** equal to the hours worked on average by each worker, being:

IWH working hours, corresponding to the hours of presence on the job determined by the collective agreement for the working period considered (average data that also considers the organization of shifts);
η maintenance labour efficiency (measured as indicated in Sect. 4.1);
a absenteeism index, estimated by statistical data.

The **Maintenance Labour Requirement** will be determined by:

$$MLR = \frac{BMAV + PMAV}{IAA}$$

5.5 General and Complementary Equipment for Production

5.5.1 General Services Equipment and Energy Consumption Optimization

Energy employed in transformation processes can be divided into:

1. **Electric Energy**, supplied to technological users through primary and secondary distribution nets; these are designed and dimensioned in relation to a unit's production and power consumption parametrically estimated, depending on types of processes and production volumes. The connections to the specific equipment are realized based on layout of machines and installed systems. Devices for monitoring consumption are opportunely positioned, along with one for safety protection and other specific needs for equipment power supply (high voltage smoothers, frequency converter…).
2. **Energetic Fluids** employed in material transformation processes: methane, propane, ethylene, steam, compressed air…
   Each one of the above fluidic vectors is supplied by plant specific nets, or through control units close to the users. Distribution nets are designed based on technological layout and in relation to power consumption parametrically estimated, applying necessary saving systems, pressure regulation and flow control. Some fluidic vectors, conversely, are supplied by generators/accumulators integrated into the technological equipment itself (for example: control unit for hydraulic or pneumatic servo-controls).
3. **Other Technological Fluids** used in the transformation process: inert gas, industrial water, liquid for lubrication and refrigeration….

For the fluidic vectors considered "critical" from a safety point of view (compressed gas, inflammable or hazardous fluids) strict safety prescriptions must be applied.

Whenever production processes are energy consumption intensive, it is appropriate to set out some modern systems for monitoring, to allow the operators to adjust the process parameters requiring necessary specific maintenance intervention quickly.

In order of effect on transformation cost, we review the automotive production processes of high specific energy consumption level below:

- body in white painting;
- foundry for cast iron or light alloys;
- polymer transformation;
- steel cold stamping;
- metal machining;
- body in white welding and assembly.

Relative to the above processes, it is proper to adopt energy-saving technological solutions. In particular, it is necessary to apply modern systems for heat generated from the recovery of transformation processes and for the recycling of rejected fluid and solid material.

Furthermore, it is to be underlined that conditioning necessary for solid functioning of some technological processes and occasionally for the needs of the working environment is a relevant part of energy consumption in automotive factories. Relative equipment must be well designed to optimize thermal flow and to be easily regulated while functioning, according to need.

Another point of attention concerns the lighting systems for the shop floor and the offices, which must be set considering energy-saving criteria as well as the systems mentioned earlier; this is to assure the necessary level of illumination in the working areas, distinguishing them from the transit areas, the warehouses and the other secondary areas, while respecting safety criteria. Lighting groups should be high performing, divided by sectors and regulated automatically by photoelectric cells with programmable thresholds.

In the future, more restrictive norms for energy-saving criteria and CO_2 emission will be applied, so as to engage General Equipment Makers in developing more sophisticated technological solutions.

5.5.2 Co-generation Equipment and Electric Energy Consumption Control Systems

At some automotive final assembly plants, where scale of economy and localized activity provide convenience, co-generation equipment for energy production (thermo-electric power stations nearby or inside the plant) are employed. In these situations, steam generated from low pressure turbines can be directly used both for heating of the working environment and for technological use, especially for body painting processes. The economic advantage is more significant when these power stations are connected to public networks and can interact, absorbing energy when more is needed or providing energy when there is an excess. In the future, photoelectric panel application will be more common, so that solar power will be used for conditioning working environments.

To optimize energy consumption, it is important to have dedicated monitoring systems. Starting from data collecting in real time, weekly or monthly reports are organized to better focus countermeasures or improve necessary actions.

It is suitable to appoint a specific person responsible for "energy-saving" activities for each production plant, with whom all unit managers should collaborate in general equipment management. Concentration of activities on certain working shifts, as well as considering the differences in energy fees, appropriate area sectioning, and modular equipment adoption, are all typical actions for limiting energy consumption.

From the moment that energy resources are "critical" by definition, it is appropriate to control equipment efficiency and plan necessary corrective actions. The analysis of final specific consumption allows for monthly monitoring of deviations from standard utilization.

In this regard, it is important to consider that:

- in Italy, electric energy cost is higher than the average in EU countries, and even more if compared to North America or Eastern Europe;
- in other world areas with poor mineral resources, but where industrial activities are developing fast (such as Brazil or China), serious electrical energy shortages can occur periodically, as a consequence of sudden hydric lacks that can condition trends in production activity.

Let us conclude this section by underlining the importance of fire fighting systems that are included in the general equipment. Maintaining efficiency within these systems, according to risk prevention policies, is the responsibility of Plant Technical System Management.

5.5.3 Auxiliary Equipment and Exhaust Material Management

By auxiliary equipment, we mean:

1. equipment for technological fluid treatment and recirculation, such as lubricant and cooling liquid for metal printing and metal machining process, industrial water for body in white pre-treatment processes…;
2. equipment for collecting and recovering chips resulting from metal transformation processes, such as foundry wastes, blanking and shearing chips…;

3. equipment for regeneration and exhaustion of solid and liquid materials resulting from transformation processes, such as sands and soil in foundry, lubricants and coolants, chemical products in the painting process;
4. equipment for decreasing dust and the depuration of gas resulting from metal and polymer transforming processes, chemical-physical treatments and surface coating and functional dynamic tests in cells for engines and vehicle.

The above types of auxiliary equipment are becoming more relevant from a technical and economic point of view, in relation to the more severe norm for internal and external environmental emissions. Managers and those responsible are in charge of assuring that these rules will be respected and that maintenance and control of this equipment will be done according to legal requirements.

Working costs for this equipment is principally due to the specific energy consumption and particularly expensive consumable materials, such as diaphragms for ionic exchangers and liquids for technical cleaning. As a consequence, the management of this equipment requires serious attention, both from a cost point of view and from a legal liability point of view.

5.6 Tools and Consumables Management

5.6.1 General Criteria

For many technological areas, the use of consumables significantly influences the exercise costs for Plant Technical Systems and their efficiency.

Let us remember that the distinction between direct material and indirect material (consumables) derives from the "transformation" concept. Those materials that are transformed and transferred to the final product are considered direct material. Both tools and tooling subject to consumption and normal substitution, other than consumable materials necessary to perform the transformation processes, without being transferred to the final product, are considered indirect material.

Management of tools and consumables includes the following activities:

1. development of specific technical characteristics, in relation to transformation processes;
2. specific consumption analysis and definition of requirements (parametric to production volumes);
3. checks on feed planning and relative acceptance;
4. planning and controlling of costs related to consumption, in relation to each centre of responsibility (productive units).

First of all, it is necessary to perform activities of applied research, exploring offers and convenience in introducing innovative solutions, as explained in Sect. 5.6.2.

If solutions and relative processes are technically delivered, consumption standards are defined. These standard values, correlated to purchasing price and maintenance costs (re-generation), are input for the planning of material supply and consumption expenses.

In the same manufacturing system, consumable material cost is assigned to each product in relation to the working cycle time per unit (WCT/K), or considering other parameters (for example, the weight of developed products in the case of metal processes).

All of this concerns the analysed cost, while for the purpose of controlled management, consumption expenses for consumable materials are normally divided in proportion to volume of hours of each cost centre (DAV, Developed Activity Volumes or, EWT, Employed Working Time, see Sect. 4.2).

5.6.2 Tools and Tooling Subject to Quick Usury

For the importance of the manufacturing of mechanical components in automotive applications, we specifically examine the topic of the management of tools used in mechanical machining.

- *Precise stamping technologies for steel components*
  They are subject to fast consumption and are key determinant factors for "process capability" and economic savings: dies and tools specific for forging and for sintering-forging processes and dies for cold extrusion and form coinage. The precision of produced elements and the over-metal entity, necessary for the finishing steps, depends on them.

The absence of structural (cracks, inclusions…) and geometric defects are determinant conditions for final product availability during functioning and for limiting machining rejects.

Determinant factors for this purpose are:

- steel conforming to technical specifications and absence of defects due to lamination and rolling processes;
- correct shearing of slugs from profile bars (weight conformity, absence of reworking);
- intrinsic "process capability" in the stamping process, resulting from process continuity and matrix/die management.

Let us deeply examine the technical aspects of dies:

1. Building designs are studied with accuracy, simulating forming processes and optimizing step by step phases through the application of CAD/CAE techniques and using materials with high structural and superficial resistance.
2. Die substitution is planned to prevent defects, searching for a more appropriate intervention frequency to assure the necessary level of "process capability", with the minimum industrial cost. Equipment technical availability (A) is greatly influenced by the life length of matrix dies and by the continuity and regularity of the productive process. Particularly important for forging processes are: temperature control for incoming machined materials, optimal deployment of progressive stamping phases. Particularly important for cold extrusion processes are: geometrical uniformity, heat treatment and regular slug lubrication, incoming in the stamping process.
3. Matrix die building should be localized near the stamping unit, because of the strict correlation between the two processes. A die building centre includes: numeric control tool machines, programmable machine for electrical erosion, measuring tools with profile projection, heat treatment equipment for steel and PVD coating.

The correct technical management of processes and dies is determinant for assuring the final quality of these steel components, which are widely applied in the "important" functional class for availability and safety of vehicles and powertrain systems.

- *Metal removal machining technologies*
  This technological area covers all manufacturing activities of mechanical components. Technical management of tools, together with equipment maintenance, is particularly important for the optimization of processes.

Let us review the following topics, considered essential for the technical management of cutting tools:

- **Material and technologies for tools construction**
  These are characterized from progressive innovations due to the introduction of ceramic coating (form tools), metal-ceramic composites (cutting inserts) and poly-crystal composites with a polymeric matrix (abrasive tools). The increase of economic cutting speed, obtainable through these innovative materials, allows for improving the productivity of machining systems dedicated to the mechanical components of powertrains and space frames.
- **Construction characteristics of standard tools and tool-carriers**
  The evolution of solutions for cutting materials has occurred simultaneous to research for suitable configurations for detaching and breaking chips, dry machining and operations for the fast retooling of insert tools.
  It is to be emphasized that materials used for form and cutting tools are used in different applications, in relation to working speed and the economic life length

of cutting tools: super-rapid steels allow for lower cutting speed, even if they are very resistant to shocks and, for this reason, are convenient for roughing and semi-finishing. Ceramic coated tools (Al_2O_3, Si_3N_4...) have a high cost per unit, but allow for higher cutting speeds at economical working conditions. Finally, poly-crystalline tools (PCD and CBN) are particularly expensive, but allow for reaching very high cutting speeds, with a contained power supply, so as to generate precise super-finishing.
Analysis of applications is based on experimental techniques and research for optimal parameters for economic efficiency, comparing benefits emerging from speed-up of the working cycle to a higher incidence of technical stoppage for tool change. This technical–economic analysis leads to the definition of optimal conditions for minimizing working cost, considering the following parameters and cost factors:

- theoretic productive capacity, equivalent to the system's hourly virtual production (HVP)
- incidence of technical stoppage for tool change, influencing process inactivity (pu)
- expenses due to tool consumption and maintenance.
  The above parameters have been defined in Chap. 3.

– **Construction characteristics of form specific tools**
These include evolving profile makers, broaches for the enlargement of shaped holes, diametric high precise enlarger, grindstones and relative revitalizing devices.
With the evolution of applied materials, construction techniques for tools have also improved and higher levels of quality and lower costs have been reached.

Considering the above technical aspects, for effective tool management, the following rules are to be observed:

1. **Set-up and managing of pre-setting activities**, consisting of preparing the tool carrier outside of the machine, adjusting the inserts precisely and verifying geometric positions of the cutting tools, in relation to cutting directions.
2. **Set-up and management of tool regeneration**, including sharpening of integral cutting inserts, regeneration and dynamic equilibration of grindstones...
3. **Organization of tool changes**, according to pre-defined programs, that consider life length of tools, evaluated on a statistical basis, for advance intervention in spite of critical consumption limits; simultaneous interventions on more tools in the same machining unit are generally organized to shorten "set-up-time", searching for the best convenience point for substitution frequency and way of intervention, to increase technical availability and decrease working costs of the systems.
4. **Controlling quality in acceptance and focusing the choosing process of tool suppliers**. For this purpose, statistical control methodologies are applied and performances are evaluated in relation to purchasing costs.

5. **Adopting more convenient lubricant and coolant fluids,** to improve tool efficiency and technical life, also considering regeneration needs and costs derived from recycling and exhaustion of consumable liquids.
6. **Optimizing working speed and tool consumption**. This activity requires teamwork, involving more job specializations: manufacturing engineering experts, tool machine specialists, tool specialists, and those responsible for productive units.

Let us conclude the topic of tool management, underlining that we have dealt with mechanical machining in more detail owing to the greater implications of this aspect. We must say that, for other technological areas, the criteria are almost the same. From the relevance that assumes in automotive applications, we also mention the following types of tools and consumables subject to high usury:

- *resistance welding electrodes*, which highly influence qualitative levels and efficiencies in spaceframe and body assembling systems;
- *filling materials,* powered by welding wire for arc and/or braze-laser welding (above considerations are almost true);
- *special steel insert and hard league filling materials,* applied on die parts subject to consumption, which highly influence quality levels and transformation costs in sheet metal printing and die casting processes.

5.7 "Total Productive Maintenance" Approach

Before concluding this chapter, it is useful to remember the organizational approach for maintenance activities known as "Total Productive Maintenance" (TPM). Its historical origins begin in 1960, when major Japanese industries, determined to compete on a worldwide market, understood that they had to improve the quality of their products dramatically; one of the weakest points was the management of plant technical systems maintenance and, to improve in this regard, they adopted a benchmarking method taken from the United States of America; the concept of "preventive maintenance", naturally adjusted to a Japanese approach and also considering the increasing level of automation in the production processes, became one of the key strategic factors of success of the Japanese industrial model.

Once more, we find Toyota to be the main applicant of this approach that was later, during the 1970s, also adopted by other Japanese enterprises. TPM approach combines professional and autonomous maintenance activities to improve equipment efficiency and establish a total productive maintenance system through the adoption of techniques based on the continuous improvement (kobetsu kaizen) methodology that we will look at in Chap. 8.

This method gave production managers the responsibility of improving *Plant Technical Systems* constantly, putting the human factor at the centre of attention. The TPM acronym stands for:

- **Total,** because it **actively involves** all employees working in the industrial processes (from production to maintenance, quality and supply chain) in the improvement of equipment efficiency.
- **Productive,** because the target is the **continuous improvement** of "manufacturing systems".
- **Maintenance**, because the goal is to **maintain efficiency and effectiveness** of the equipment during use.

TPM is oriented to the "excellence" of production processes and the ambitious target of zero breakdowns. For this purpose, some technical-managerial topics are systematically approached, such as:

- *improving working cycle* in every phase, to speed-up, stabilize and make more continuous the production flow in every stage;
- *improving the Overall Equipment Efficiency (OEE), increasing* the available production capacity;
- developing techniques for *preventive maintenance*, assuring its application during the entire system cycle;
- developing a full *interfunctional integration* between Product Engineering and Process Engineering, Manufacturing, Maintenance and Quality Departments;
- obtaining the *involvement* of all workers in the productive and maintenance process, to optimize functioning costs (minimum indirect labour, minimum consumption of consumables and energy, maximum utilization of investments);
- improving *safety* in utilization of equipment;
- reducing *safety stocks* between the equipment;
- promoting *continuous improvement* initiatives for increase in quality and productivity through "teamwork" oriented to a "problem-solving" approach.

In Chap. 8, we will see that the TPM approach, together with those of Total Quality Management (TQM), Total Industrial Engineering (TIE) and Just in Time (JIT), are the main pillars for a competitive "Production System", referring to the Toyota model. In the 1990s, many western enterprises began to study the Toyota Production System as an ideal benchmark.

Chapter 6
Logistics and Supply Chain Basics for Automotive Application

6.1 Historical Evolution of Logistics and Actual Strategies

Logistics started with ***military applications***, as a function for supplying material and equipment necessary for war missions, getting them to the right place at the right time for an operation's success, and assuring a connection between the front and back lines of the army. In ancient history, the Roman Empire succeeded in extending and protecting its extended borders for a long period of time thanks to advanced military logistic organization, with the ability to transmit information and move legions very quickly for that time.

In modern history, at the beginning of the Nineteenth century, Napoleon Bonaparte rediscovered the importance of strategic Logistics, making it a very important pillar in the organizational structure of his army. As a consequence, all of the top military academies introduced Logistics into their war science departments.

Still focusing on military applications, it is very interesting to observe the way logistic efficiency was a determinant in the final result of the Second World War. This is particularly evident in the disembarkation of Anglo-American forces at Normandy, the utilization of American Navy fighting planes through aircraft carriers in Pacific battles, and the ability to encode radio-transmitted messages while decoding those of the enemy. These strategic-logistic prerogatives of the Anglo-American forces made it possible to outmaneuver German and Japanese armies, despite the latter being superior in combat.

In ***civil applications***, Logistics only assumed an important role beginning in 1960, in particular for public transportation and the delivery of goods. Only through the application of suitable logistic and information technology systems has it been possible to develop major airports and harbors. In the same way, an enterprise's success in major distribution of goods has been a consequence of the application of effective logistic systems.

In ***industrial manufacturing applications***, the "automotive industry" has been the first to develop logistic and information technology systems to manage:

M. Gobetto, *Operations Management in Automotive Industries*,
Springer Series in Advanced Manufacturing, DOI: 10.1007/978-94-007-7593-0_6,

- numerous and technologically complex products;
- production activities at more plants, linked together through multi-purpose transport systems;
- sales and after sales branched networks, geared towards guaranteeing good service.

Let us take a look at the main historical periods in which the most significant changes for logistic systems occurred in the automotive industry.

6.1.1 1925–1975: Logistics as "Material Management" Support

Through the "products diversification and production activities delocalization" policy adopted by A.P. Sloan in the United States during the '20s, "material management" had to approach more complex problems, compared to those of the previous era. "Material handling" studies were placed beside manufacturing engineering studies of a Tayloristic origin. Means of transport were defined on a standard basis, as were the equipment and buildings dedicated to inventory and material handling.

During the '60s, logistic processes improved principally for second generation information technology systems, developed by specialized companies begun in North America and grown worldwide (IBM, HP…). Data electronic elaboration allowed for sped up planning and control processes. Nevertheless, at the time, transmission networks were still based on analogic techniques and did not allow for the exchange of real time information.

Manufacturing systems operate at a high rhythm, but are not flexible; changeover requires long "set-up-time". Production is normally set by economic batches, and the logistic process assures continuity of production only through high levels of stock and long lead time at the plant level. The work in progress includes a significant portion of the financial resources and there is still considerable inertia in answering commercial variation requests.

This logistic model was revealed to be inadequate when two important oil crises occurred in 1973 and 1979, reflecting negatively on market trends and transport costs and generating high demand instability. Furthermore, an increase in oil and energy costs resulted in severe consequences for inflation, the consequent cost increases obliging carmakers to reduce their working capital dramatically.

Top-management realized they would need to reform logistic processes deeply and turn to organizational consultancy specializing in the introduction of new systems, especially for the purpose of overcoming internal resistance to changes in management.

6.1.2 1975–1990: Logistics Oriented to "Time-to-Market" and "Commercial Network Service"

During the second half of the '70s, "New Generation's Information Technologies Systems" became available; these systems had applied software models, specifically developed for the planning of operations and management of supply processes (Product Data Management systems, Material Requirement Planning...). Big Information Technology Enterprises offered interesting and convenient solutions for investments with an advantage breakeven point, thanks to a reduction in people employed in data collection and input processes, paper reports consulting processes, and so on. With automation, the availability of logistic processes also increased and production flows were under control in real time, especially with the introduction of "PC-based" peripheral information technology systems.

At the same time in Japan, Taiichi Ohno of Toyota was introducing a visual management system for material supply, which worked in a kind of **pull** logic for application of the Just in Time principle called "kan-ban" (tag), a simple and effective method, as seen in the section on basic concepts.

Productive flow organization is definitively oriented to a pull logic, especially for final assembly of cars and big components. Production is based more on customer orders than on forecasts.

Nevertheless, with WEB networks still not widely spread out, it was not possible to extend information technology systems along the "supply chain", operating interactively and with direct connection to the dealer's network. So, during the '80s, logistic processes improved at the plant and delivery levels, but production and material supply planning was still performed in cascade along the supply chain, with a monthly re-order system. At any rate, production flow appeared leaner compared to the past, and lead times in reacting to customer demand changes were shorter, also thanks to the diffusion of programmable manufacturing systems (numeric control machines, robots...).

It is important to consider that, even in this historical period, "time to market" and "customer service levels" were recognized as strategic factors for an enterprise's success. Logistics focused on these new important targets.

6.1.3 Actual period: Logistics as an "Integrated Process" to Support "Supply Chain Management"

As seen in Chap. 1, major "carmakers" adopt product strategies based on the concept of a "world car", with "global sourcing" purchasing policies (that we will analyse in Chap. 7). Many industrial co-operations through "co-makership" are signed between makers of powertrain systems and carmakers. So, it becomes very important to connect the different worldwide productive sites through an integrated information technology system, assuring homogeneous product from a technical point of view and homogeneous logistic procedures. Integrated Product Data Management Systems, as described in Chap. 2, can easily accomplish this task.

During the '90s, the availability of ICT systems and development of the WEB network allowed for a new deep change in logistic processes.

In conclusion, we observe that, according to new trends, Logistics is not only a specific company department, but an interfunctional integrated process that supports Supply Chain Management.

Let us conclude this short historical overview of Logistics, emphasizing that it has always been influenced by information technology and communication systems progress. Nevertheless, although much evolution has occurred, the primary target is still ***providing materials in the right quantity, at the required quality, at the right point of use, in the right time and with the minimum cost***.

By this definition, "right" means what is established between Customer and Supplier through the several stages of the "supply chain", in relation to final customer expectations and the commitment assumed by the producer. Minimum cost means having an economic global logistic process, including: dedicated system working cost, committed services cost, and financial burdens due to work in progress capital (production and distribution).

To the reader that still does not know production planning and inventory management criteria, we suggest looking at the final section, on basic concepts, at the end of this book. It deals with general basic concepts not specifically related to automotive logistic processes that, conversely, are treated as such in this section.

6.2 Logistic Flow in the Supply Chain

The modern concept of *extended enterprise* involves the whole set of companies working for the development of a specific product line or services offered to customers or users. In an automotive compartment, the enterprise that owns the "brand" is normally also the Final Producer and operates as the figurehead in business management for the market; other companies belonging to the same industrial corporate entity or external suppliers that cooperate directly or indirectly in these industrial and commercial activities also participate in this success, creating the so-called "supply chain".

According to this logic, we define all company units that participate in the productive and delivery processes as **Supply Chain (SC)**. In an automotive organization, commercial relationships and contracts of cooperation are normally managed by the following departments:

- *Marketing & Brand Management*, for Dealer relations;
- *Purchasing Management & Procurement*, for Supplier relations;
- *Sales Operation Management*, for Customer Enterprise relations.

Methodologies for *business to business* supplies will be examined in Chap. 7, giving evidence of the role of *Purchasing & Procurement* departments.

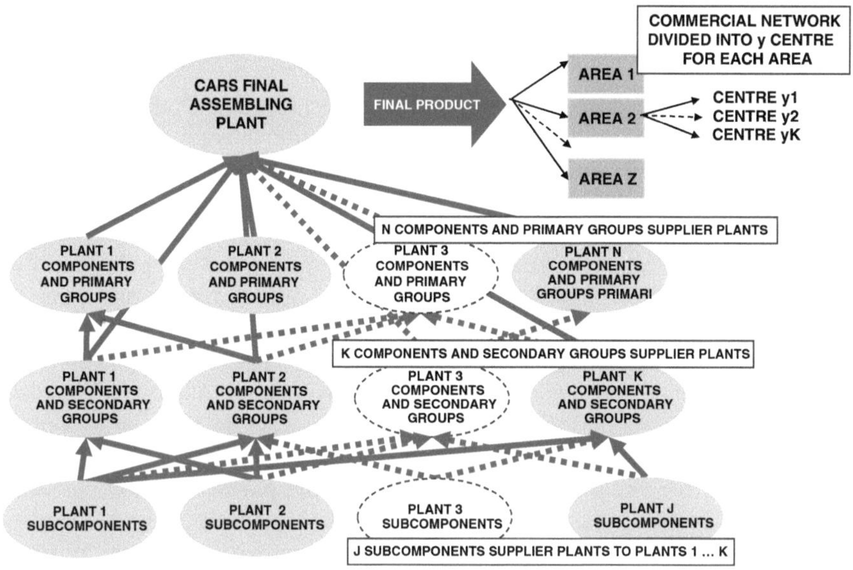

Fig. 6.1 Supply Chain structure in automotive industries

Conversely, logistic-industrial relationships are managed by *Supply Chain Management* and *Manufacturing Management* departments, following the procedures described in this chapter.

The typical SC configuration for the car business is represented in the chart in Fig. 6.1. In it, we find the plants but not the companies, as we similarly find the sites of the dealers for receiving and dispatching to customers but not the agencies for sales. It is also important to consider that the suppliers themselves operate with more production plants, often dedicated to more customers.

As demonstrated in the scheme, the SC is a tree structure, from a conceptual point of view very close to the Product Breakdown Structure or to the Bill of Material seen in Chap. 2, even if it is not necessarily the same.

- Final producer plant, or co-producer plants, is at the top of the SC (**level 0**).
- Logically linked to them, even if from far away, are the plants producing mechanical sub-assemblies and main body components and modules (**level 1**).
- Then follow **levels 2 and 3**, which include plants producing minor components and parts that can also be of high technological content, but not presenting constraints for the manufacturing *lead time* of the final product.

At each one of the above levels, we also find raw materials and standard component suppliers, which can be assimilated to *commodities*.

In the chapter on Basic Concepts, we will see, in general terms, what are the possible organizational types of logic applicable for productive flow management. Now, we refer to production activities for automotive industries and related logistic processes along the SC.

In Final Assembly plants, with reference to the block diagrams in Sect. 1.6, production flows are organized through a ***pull*** logic, considering the number of models and different versions offered to customers. It is the classic condition in which production is led by customer demand, associated first of all with vehicles and body assembly systems, but can also be transferred to engines, transmission, suspension and assembly of the main body's modules. The same logic is also true for industrial vehicles, agricultural and construction equipment applications, motorbikes and other technologically complex products. In the following section, we will look at the *material handling* criteria necessary to set the pull logic.

In remaining technological areas along the SC, other criteria are normally applied that tend more towards a ***push*** logic. In particular, for raw material and standard components that can be supplied by different alternative sources (even those geographically far away), some localized buffers are admitted. This production is normally set through logistics of economic batches, with delivery plans being dependent on transportation methods.

A mixed "***push and pull***" logic is applied for the production of components with medium–high logistic complexity; this production uses homogeneous manufacturing systems to supply a wide range of customers (including spare parts). Some examples are:

- bulky electrical and fluidic-thermal components;
- wheel rims complete with tires;
- very complex and specific castings for the manufacture of engines and transmissions.

To limit lead time of orders for final products (vehicles and powertrain systems), these upstream productive processes must be set in advance to a commercial network order acquisition process. Related supplies of materials are made in general on a sales forecast according to MPS (Master Production Scheduling), through which requirements are aggregated in homogeneous families of products, with time horizons that consider the lead time of both processes, both transportation and dispatching methods. For these types of components, carmakers normally require their suppliers to manage their warehouses nearby the final assembly plants with stock on consignment, to allow for the application of a just in time (JIT) logic in the general assembly lines.

To feed components to general assembly lines in a JIT logic, *kanban* techniques are used; kanban works to replace empty units with full ones through specific tags that travel together along the loading unit (the modern ones with automatic reading procedure) and on which are reported certain important information such as part number, utilization point… This tool allows for the simplification of productive and logistic flows, minimizing material stocks at all levels and reducing material handling operations in final assembly; in this way order lead time for final customers can be shortened.

As we will see in Sect. 6.4, automotive operational planning is focused on the first two levels of the SC (0 and 1), the production flows of which are directly dictated by commercial demand, based on collection of customer orders.

Nevertheless, the logistic process also invests the other levels of the SC and includes the following macro flows:

- Physical Flow

This begins with raw material supply, crosses all stages of transformation along the SC, and also includes stocking and handling phases up to final assembly and final testing of vehicles, from which follows commercial dispatching and distribution.

Speeding up physical flow allows for the quick answering of customer demand and short order lead time.

- Information Flow

This goes in the opposite direction of physical flow, starting from customer demand to suppliers through different stages of the Master Production Scheduling (*MPS*), the Material Requirement Planning (*MRP*) and the Production Scheduling (*scheduling*), up to the dispatching of goods (*dispatching*).

Information flow allows for constant planning and controlling of the progress of production and delivery, at every stage of SC.

- Financial Flows

This is a consequence of physical flow and is evaluated based on data detected from the information flow at every delivery point incoming and outgoing at the different stages of the SC. Direct materials are transformed and aggregated, progressively adding value, up to product completion and customer delivery.

It is fundamental to observe that, under the same economic conditions (purchasing prices, transformation costs), working capital reduces in inverse proportion to physical flow speed Fig. 6.2.

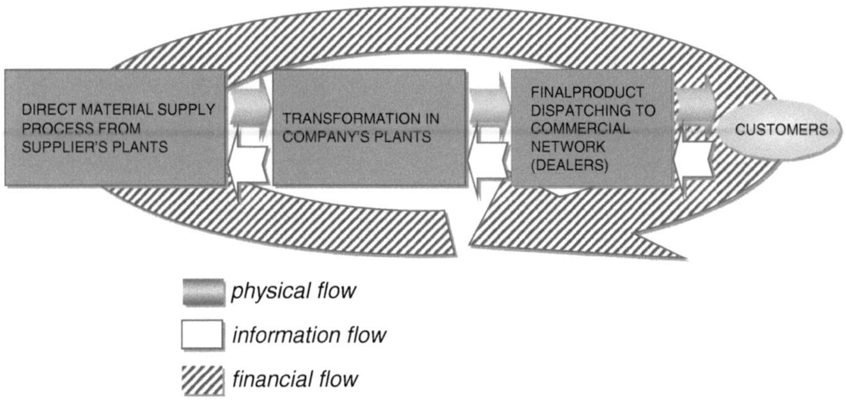

Fig. 6.2 The three main flows of logistics

6.3 Material Handling and Inventory Management Methodologies

Operations and means necessary to **transfer, stock and deliver** goods are called a ***material handling system and*** involve the following areas of the integrated logistic process:

A. Handling between the plants along the Supply Chain
B. Plant internal handling
C. Handling for Commercial Delivery

As seen for manufacturing systems in Chap. 3, to set competitive material handling systems, it is necessary to analyse operations deeply and choose more convenient solutions. For this purpose, a "*material handling engineering plan*" has to be set, following the logic shown in the chart in Fig. 6.3.

Based on *make or buy* policies adopted by automotive industries, activities of type A and C are charged to companies specializing in the transportation and delivery of goods, operating, of course, at international levels.

Area C is particularly important, because an efficient system of transport and delivery of cars to the commercial network contributes in a significant

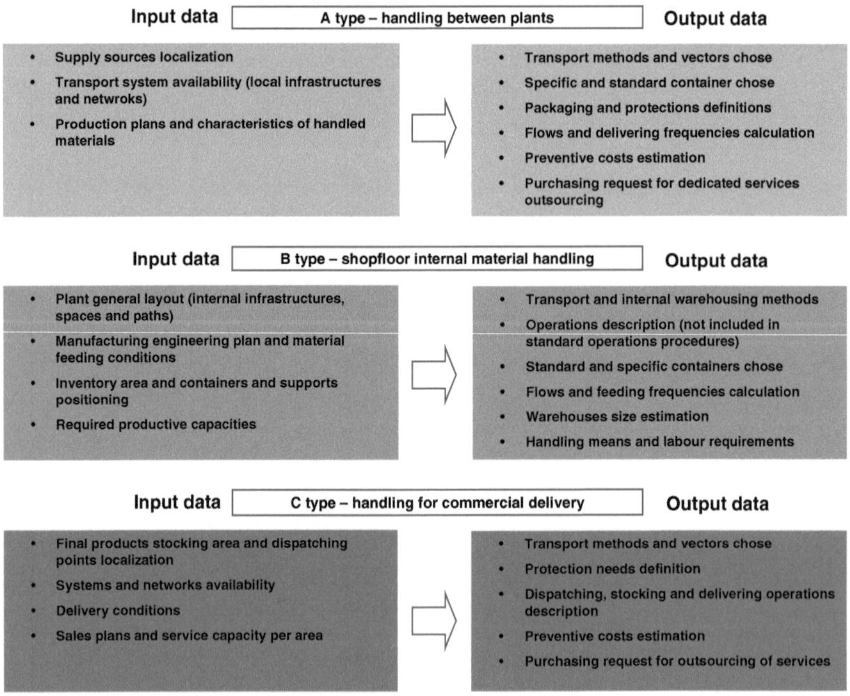

Fig. 6.3 Material handling engineering input and output data

way to reducing the lead time of orders. This area also includes operations of supply, stocking and delivery of spare parts necessary for after sales technical assistance (important operations for the customer service level and for the company's final net operative result).

Topics related to transport systems both external and internal to the shop floor (Areas A and C) are not addressed here. We note only that in the general layout design of a plant, qualitative and quantitative incoming material and outgoing product flows are estimated. As a consequence, infrastructures necessary for connecting the plant to suppliers and commercial networks are considered.

Now, we focus on the internal handling operations (Area B), which are typically managed by the factories, with rare exceptions. Material handling engineering is oriented towards searching for the best compromise between the two following opposing needs:

- full utilization of transport vectors operating in Area A, to minimize working costs;
- high frequency for feeding of production line, to operate with a lean flow according to "just in time".

For this purpose, the criteria for choosing transport means and calculating load sizes are:

(1) aggregating parts in container elements, calculating the size of ***loading units,*** so that they are transferable under safe conditions by using normal means of handling, maintaining the same configuration during all handling operations along the logistic process and allocating containers directly at utilization points;
(2) choosing ***types of containers*** that will assure protection of goods and make withdrawal and warehousing easier, according to standard operation instructions and;
(3) organizing ***delivery batches*** by joining more loading units, so that transport vectors can be well used, even through mixed deliveries (more element designs, collected and delivered according to planned transport paths).

Obviously, the size of the transported load for each mission will determine how low the cost of the transported unit will be, without considering other factors.

Handling systems used to feed direct materials into the productive process can be classified as:

1. ***Supports/Primary Containers***, which sustain parts and group them together both in the transferal and warehousing phases; they are studied to save space and can be divided into the following types:

   - *standard pallet,* on which the single parts or secondary containers are placed on top (if necessary, blocked for safety reasons through specific packaging solutions);
   - *standard containers with fixed sides (bud), with more layers of parts and eventual protective separators;*

- *special containers with collapsing sides, with intermediate layers engineered for positioning of parts.*

These supports are of a standard modular shape and can be handled by forklift and stocked one on top of the other to save space in the warehouse. They are built of steel plate and are strong enough to be used and re-used over a long period of time and a wide range of transportation. Conversely, ***pre-printed inserts*** are specific (normally built in reinforced resin), used for protecting elements from shocks and collision during handling.

For smaller elements, ***secondary containers*** are used (bins, standard or specific baskets), able to be handled manually and put on the standard pallets. Where necessary, protective films or recycling plastics or cartons, the utilization of which generates cost for exhaustion, are used for packaging.

Special containers with movable sides are subject to time-based maintenance, because deterioration can generate severe quality defects in the production process.

It is necessary to check incoming and outgoing flows for each of the above recyclable supports, at every stage of the SC, to equilibrate interchangeability and finalize destinations.

2. ***Big containers***, accessible through forklifts, inside of which are the above-mentioned primary containers, composing the loading unit.

These are universally used, both for multiple trans-boarding on transportation platforms or outdoor warehousing on factory loading docks.

3. ***Internal transport vectors***, used for transfer of loads and elements, inside the plants, which can be classified in relation to the different ways of driving:

   - *manual vectors,* such as: vehicles, small ro–ro carts, trans-elevator carts, trans-elevator equipment with fixed structures (cranes, hoist, …);
   - *automatic vectors,* such as: continuous or power and free electro-mechanical transport systems, auto-motors cart with monorail or guided on floor tracks, programmed trans-elevator equipment, handling robot…

We will refer now specifically to final assembly lines for vehicles and mechanical groups, characterized by a high logistic complexity. Supply flow must correspond to the operative speed of the productive process, which is the hourly virtual productivity HVP, referring to each assembled component (k) and to the relative utilization coefficient (u), according to the bill of material.

Even though the decision as to the flow of all components/material to be fed into the general assembly line is one of the key activities in affording the success of a lean manufacturing system, we need first to clarify the concept of logistic complexity and a first important output that results from it.

We have already stated in Chap. 2 how technologically complex a vehicle is, referring to the bill of material, and in that section, we noted that this important document is also used to manage the material requirement planning process that we will approach later on in more detail. In a bill of material for mass production

Fig. 6.4 A car passenger vehicle main component deployment

of small cars, we have more or less 4,000 components with the potential to become more than 15,000 if we consider all the variants in options given to the customers.

As such, from a logistic point of view, we first have to consider that for all of those options, each component has to be delivered to the final assembly plants in the best condition for assembly and eventually managed in internal warehouses, transported to the final assembly line and finally mounted on the vehicles to become part of the product. So far, this is giving us a rough idea of what logistic complexity means Fig. 6.4.

Practically, every piece (component or raw material purchased) to be delivered to a production process has got its own given **Intrinsic Logistic Complexity**, listed in order of importance:

1. the **cost** of the item: we have to consider that the more an item costs, the more the value of the inventory we will have to manage will increase and, from an economic point of view, only when it is part of the finished vehicle will we be able to see a return on that value by selling the car to the customers; practically, we are saying that the more expensive items are very complex to manage from a logistic point of view, since we should have the very minimum level of them in our internal processes, theoretically only those assembled on the vehicles, to minimize the cost of production
2. the **physical characteristics** of the item: physical dimensions, volume, weight, fragility, damageability and features that will eventually expire can determine special care from a logistic standpoint; let us think about the extra usage of space, physical efforts required for handling or specific handling means, usage of specific packaging, etc.; so far, when we have an item with these

characteristics, we need to consider that stocking and handling may become hard to manage
3. the **number of variations** of every single item: if the vehicle configurations permit special options for the customer, it could be that, for every single supply, we have a large number of variations (think about seats, wiring harnesses, wheels, cockpit modules…), and this could require extra-space usage on the final assembly lines, as well as in the warehouse, and could also oblige operators to engage in many NVAA (Non Value Added Activities, specifically walking).

So far, combining those three criteria, it is possible to assess the logistic complexity for every item included in the bill of material and assign a ranking to each of them. This exercise leads to **material classification**, which represents the starting point for every decision made about the ideal material handling flows for every component in the final assembly process. We can imagine that an expensive, bulky item, with many variations, such as the seats, will stay at the top of the classification, while small, cheap and common nuts and bolts will occupy the last positions; thus, ideally, the seats will not be stored in our warehouse but directly supplied to the assembly workstation, while the bolts could also be managed from an internal warehouse to the assembly line.

We will reconsider this concept in Chap. 7, when we will examine global sourcing policies. In the following table, there is an example of how to create a material classification starting from the bill of material and how it is possible from a material classification to associate ideal flows with a single category of components. What is important to underline is that, while the first is intrinsic to the product (it is part of logistic complexity), the second is a logistic decision.

Table 6.1 Material classification example

Class		Part typology	Sub class	Sub group	Examples
A	A	Expensive	AA.1	Bulky and many Variations	Engines, axles, dashboards, gearboxes
			AA.2	Bulky	Side panels, spoilers, catalytic converter
			AA.3	Many variations	Junction boxes
			AA.4	Others (monodesign)	Navigator system
	B	Bulky	AB.1	Many variations	Frames, cross members, wheels, noise insulation panels, tanks
			AB.2	Others (monodesign)	Snorkel, doors
	C	Many variations	AC	Other	Engine supports, hoses, small Pipes, mirrors
B		Normal	B.1	High rotation	Light, silent block
			B.2	Low rotation	Rotating light
C		Small and cheap	C	Other	Screws, bolts, nuts

In the example of material classification displayed in Table 6.1, starting from the cost criteria and going through the other two, each item of the bill of material can be assessed and linked first to a specific class and then to a sub-class. Once we have identified all classes and sub-classes, it is possible to identify an ideal way of delivering the specific sub-classes to the line in order to follow the main guidelines for internal handling:

- In sequence flow
- Direct Flow: Just in Time by use of kanban card systems
- Indirect Flow: feeding from internal buffers or warehouses.

For concerns of external handling, a good match between the ideal delivery flow and the supplier's geographical location will depend mostly on the global sourcing policies adopted by the company, a job for the Purchasing Department in collaboration with Logistics and the Supply Chain. This aspect will be examined in more detail in Chap. 7.

Once we know the ideal policies for feeding parts and components to the process, we have to determine the proper **handling unit** (the quantity of elements/goods we have to handle during a single handling) and the **feeding frequency** of each of them. For each element k (expressed in number of missions per hour), the frequency is determined by the ratio:

$\mathbf{FF_k = (u \cdot HVP)_k/U_k}$, where $\mathbf{U_k}$ is the number of elements included in the handling unit.

The calculation of $\mathbf{U_k}$ should be consistent with the withdrawal condition of spaces and elements as foreseen in the technological layout.

The number of transport missions necessary to feed the manufacturing systems with continuity, within a certain period of working time (working shift, week…), corresponds to:

$\mathbf{OEE \cdot FF_k \cdot PWT/N}$, where:

PWT is the system's planned working time (hours)
OEE is the system's overall equipment efficiency
N is the number of handling units transported simultaneously, in relation to the capacity of carriers employed and the composition of transport of economic batches.

Criteria applied for inventory management corresponds to the logistics mentioned in the section on "basic concepts" at the end of this book (deterministic and stochastic/probability methods).

For this, it is fundamental to distinguish the utilization of intermediate stocks in the productive process:

(a) *stocks necessary for* ***production by batches***, the management of which is based on economic batch or minimum sustainable batch policies, according to what is laid out in the section on "basic concepts";
(b) *stocks necessary to* ***make two independent process stages***, for compensating for differences in flow speed and limit interferences due to technical failures (first-in/first-out dynamic warehouses, known as *dynamic buffers*);

MH SYSTEMS TYPOLOGIES / APPLICATIONS		Incoming goods unloading and warehousing	Manufacturing systems feeding	Inter-operational connections	Final product warehousing and dispatching	Tools change
Manual handling systems	MOBILE TRANS-ELEVATORS	□	X		□	△
	RO-RO CART TRAINS		X			
	FIXED STRUCTURE TRANS-ELEVATORS	X			X	X
«Hard Automation» systems	CONTINUOUS TRANSPORTERS		X	X		
	«POWER AND FREE» SYSTEMS		X	●		
	SEQUENTIAL AUTOMATIC VEHICLES		X	●		
Automatic programmable systems	AUTOMATIC GUIDED VEHICLES		△	△		
	NUMERIC CONTROL TRANS-ELEVATORS		△	○		
	ROBOT			△		

Function symbols

X fixed cycle transfers

△ programmable cycle transfers

● simple dynamic warehousing (FIFO logics intermediate stocks)

○ selective dynamic warehousing (intermediate stocks with flow re-sequencing)

□ transfer and warehousing distributed

Fig. 6.5 Shopfloor material handling delivery systems classification

(c) *stocks necessary to* ***set sequences*** *in mixed delivery,* for machining and assembly (selective dynamic warehouses).

A particularly relevant function for *material handling* systems is related to the goods under financial control. In the following section, we will briefly describe operations dedicated to this management:

- *goods received*: incoming loads are received, goods are identified and relative containers are set in a preliminary inventory area;
- *goods accepted:* this is done in parallel with the receiving operation, verifying documentation of loads and operating statistic controls for administrative accountability;
- *warehousing*: goods are stocked in a specific area, assuring quick withdrawal times;
- *goods preservation*: goods are warehoused with specific protection during all necessary time;
- *order withdrawal*: the group of goods composing a single order are withdrawn, verified and gathered in loading units, in relation to dispatch;
- *dispatching pre-set*: goods are packaged, applying necessary protections to assure integrity of the load during transport and elaborating necessary financial/administrative documentation;
- *dispatching*: necessary handling for final transportation loading.

Registrations done during receiving and dispatching generate active and passive invoicing, determining the economic evolution of warehouse inventories.

In Fig. 6.5, a resuming scheme for material handling systems on the shop floor is represented, in relation to logistic functions performed. Even if we don't consider building characteristics for equipment (a specific topic of the Industrial Equipment module), the chart matches *material handling* system typologies in every application to the specific logistic function required in a matrix concept.

This scheme can be very useful for the setting-up of *material handling* systems on the shop floor.

We remember that, in Sect. 4.1, we considered the workers involved in warehouse handling (indirect labour) as auxiliary functions. The relative requirement is determined based on the frequency and length of material handling operations, also considering features of handling means used.

Total *material handling* cost, in the same integrated logistic process, includes:

1. dedicated labour cost, except that which is still included in standard operation descriptions and accounted in the direct labour standard time (ST);
2. functioning cost of handling means, including depreciations;
3. cost of dedicated information technology systems, including fees for functioning and depreciations;
4. consumables for packaging and cost of protection of goods;
5. burdens derived from working capital of the logistic process.

During project setting and outsourcing of services, solutions should be sought that minimize the global cost mentioned above, considering the same functionality of systems (productive and delivering capacity, attended service level...).

Considering the whole logistic process along the SC, including delivery to commercial networks, the global cost for *material handling* generally ranges from 6 to 9 % of final product cost (passenger cars and commercial vehicles). Its incidence is relevant and should be controlled with rigour. It is obviously higher when production is far from the final products' market destinations and when the infrastructure of an area in which an enterprise operates is not suitable for the transport of goods.

6.4 Production and Delivery Planning

Here we link to what was discussed in Chap. 1 on the topic of strategic planning. For automotive industries, information resulting from strategic planning is taken up again in the following documents of primary relevance for business:

- **product range plan** and its evolution in the medium-long term;
- **market mission and targets,** referring to the different geographical areas and to product lines considered strategic for business, in relation to a *brand*'s mission (market shares and sales forecast);

- **industrial initiatives,** delivered for development of new products and for setting new production capacities needed, in relation to the above market targets and return of investments.

In Chaps. 2 and 3, we specifically examined the methodological criteria applied for developing new products and processes and new manufacturing systems at available productive sites, according to *make or buy* policies agreed to by the company.

Furthermore, to organize production and commercial delivery, a "medium term operative plan" is needed, normally on a three year horizon, determining:

(1) a catalogue of products to be delivered to markets, with reference to technical contents and margin targets (sale prices, without considering the margins of intermediaries);
(2) periods of product availability, with particular reference to launch plans for new models;
(3) sales forecast for each model, considering all the types of engine and transmission and special versions requiring specific production capacity for tool conditioning.

To realize the necessary capacities for production (manufacturing system development) and commercial considerations (sales network development) on time, specific investment initiatives are run, following the methodological criteria mentioned in Sect. 3.1.

The above-mentioned *medium term operative plan* is normally updated on a yearly basis (preliminary to the budget), after some research cycles on demand and competitive scenarios to define operative solutions, subject to delivery by Top Management. At the cycle of every year, working activity levels are updated for the three years of horizon and, as a consequence, company resource requirements are updated too (workforce, investments, purchasing needs).

Production and commercialization planning begins with the medium term operative plan and becomes more defined in the short term in each production and commercial department (month by month, week by week, day by day…) to establish requirements. These obligations are for:

1. sales managers and intermediates, who are engaged in obtaining orders from customers and delivering products on time;
2. plant managers, who are engaged in manufacturing products in the right quantities at the right time;
3. supply chain managers, who are engaged in the punctual transfer of products to commercial dispatching centres, including spare parts for after sales technical assistance.

Before describing the elaboration process for the **Operative Plan**, herein referred to as the **OP**, we have to lay out a few facts:

- OP normally covers a temporal horizon of at least twelve months (some companies cover eighteen months, to have enough information for the annual budget in advance);
- OP is updated monthly (rolling system), so as to introduce timely variations due to market trends, sales results and eventual delays or anticipations in product delivery;
- OP determines labour requirements, start of working time and monthly supply plan for direct materials, according to MRP procedures;
- OP determines shop floor *scheduling* and relative supply frequencies per day/week for direct materials;
- OP determines lead times for delivery of final product to customers;
- OP determines activity levels to be considered for annual budget and its revision every three months.

To elaborate OP, we have to start from the available catalogue for sale products, month by month. This catalogue is, opportunely encoded, in correlation with the PDM system (see Chap. 2) and represents the **product grid**, because it defines and links in a matrix the following data related to vehicles sold:

- models and relative standard contents;
- alternative solutions for engines and transmissions, eventually solutions for 4WD;
- special solutions for space frame and body, also in relation to specific markets (right hand drive, stronger suspension, cold country versions…).

For the cars, we can also distinguish:

- specific body versions derived from the basic model (2- or 4-doors, sports versions, pick-up, wagon, open top…);
- special non-standard versions that are subject to temporal and quantity constraints (special acclimatization solutions, automatic seats, special info systems…);
- other specific versions not subject to specific quantity constraints (external colours, internal dressing, wheels, special tires…);

For commercial and multipurpose vehicles, more alternative versions are foreseen for the same basic model, such as: minibus, cabin van or multi van with different size and wheel axle positions. Solutions for special use, subject to specific manufacturing lead times, are offered as well because they are manufactured by external suppliers:

- special refrigerator vans, special tiltable vans…
- special versions for transport or craft (military vehicles, ambulances, camper…);

The above shows the complexity of a product catalogue for cars, commercial and multipurpose vehicles, and that which customers can choose at order, obtaining engagement on delivery lead time.

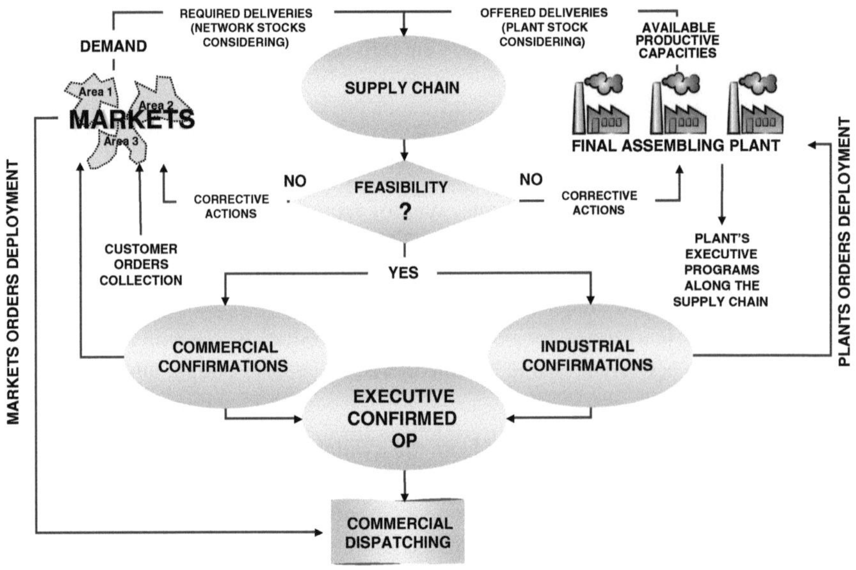

Fig. 6.6 Operative plan flow diagram

Both for total productive capacity, at a basic model level, and for capacity related to specific engines and more complex versions, there are some temporal and capacity constraints, as a consequence of investment plans and specific lead times. To test the feasibility of programs, it is necessary to aggregate sales forecasts in this way, using the PDM system discussed in Chap. 2.

In Fig. 6.6, OP preparation flow is represented, beginning at commercial demand, performing necessary feasibility tests and matching demand with an offer, up to defining the executive OP that generates manufacturing and delivery operations of the final and intermediate products at every stage of the SC.

The OP cycle operates simultaneously on two different sides:

- the commercial side, towards the different market areas where company brands are set;
- the industrial side, including final assembly plants and SC first level plants.

Coordination is made with the Supply Chain Department, who must match the evolving requirements coming from the commercial side with the evolving productive capacities offered from the industrial side; a more productive, more flexible and faster system is the answer to required variations. At the end of these feasibility verifications, production planning is aligned with commercial planning, attempting to maintain a physiological level for the finished product stocks at the plant and commercial network levels (the latter are normally charged to the dealers).

In Sects. 3.8 and 4.1, we saw how to adjust shop floor activity levels to accommodate trends in market demand, setting extra work hours when demand is higher than productive capacity or reducing working time when demand is lower. With these types of countermeasures, we can:

- contain waiting times for customers within acceptable values;
- avoid workforce unsaturation on the shop floor;
- contain the stocks of finished vehicles produced according to forecasts but not covered by customers' orders, within acceptable limits from a financial point of view.

The process of collecting and aggregating orders is coordinated from Operative Marketing Management, who are also in charge of correcting eventual imbalances between the market and single dealers. In this way, dispatch programs are balanced, transferring shares of demand from one market area to another, considering constraints due to different versions. In implementing these changes, it is important to give priority to market areas and products with higher profitability.

To make commercial demand agree with productive capacity, it is necessary to join demands by basic model coming from different market areas (meaning by body and vehicle assembly lines), and by engine and vehicle versions considered more difficult for manufacturing lead time.

The final document derived (recurrent monthly OP) distinguishes information and obligations, in relation to the temporal horizon, N being the actual month:

(1) For the medium-long term, over N + 6 months, predicted information focuses on basic models and engine families, joined according to logistics of the production lines. The mix of models and versions is estimated on a statistical basis, referring to the previous months or to a forecast made by dealers. Based on these forecasts, productive capacities and resource requirements, for plants and along the SC, are planned.

(2) For the medium-short term, from N + 3 to N + 6, predicted information is more analytical and deals with bodies and specific space frame versions and configurations considered more difficult from a lead time point of view. As a consequence, obligations for supply of less easily accessible components are taken and required weekly working shifts are planned to correspond to the market demand, assuring probable delivery time month by month.

(3) For the short term, from N + 1 to N + 2, information determines the programs' execution, because it derives from customers' orders, and eventually integrates with anonymous orders made by dealers at their own showrooms. Eventual deficiencies in specific orders or delays in the delivery of product promised for month N or N + 1 represent a severe failure for the logistic process of the company, because they can affect customer satisfaction level and cause losses in the utilization of resources (overproduction and overstock at the plant). These failures should not occur (with the exception of rare situations), because the rolling logic of OP leads to a monthly alignment of the plan, considering commercial trends and eventual constraints coming out of the production and supply processes.

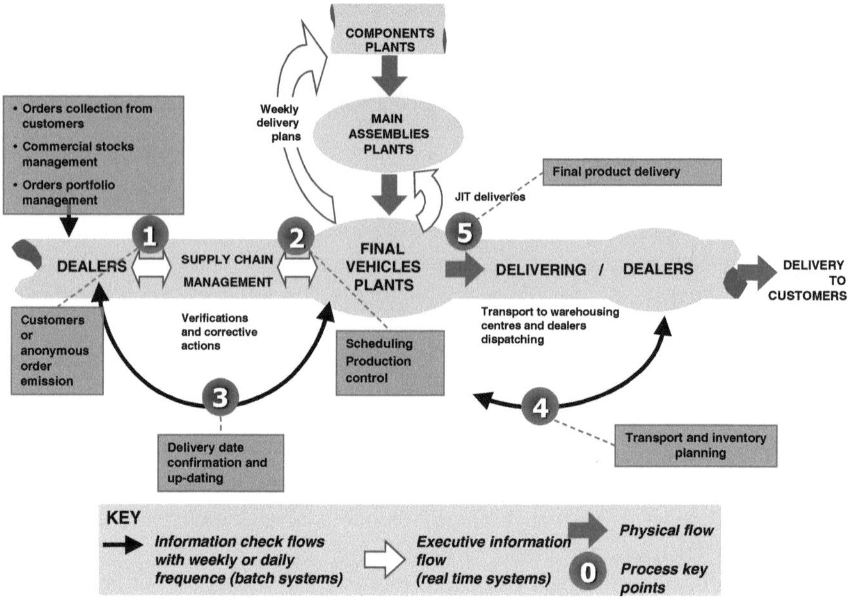

Fig. 6.7 Delivery and production planning flows

In the scheme in Fig. 6.7, the operative planning process applied to production and commercial delivery is summarized, distinguishing executive phases from those of setting and checking, necessary to set resources for adjusting ongoing programs.

Examining the phases shown above, it is evident that a core coordination is made through the Supply Chain, connecting dealers with final assembly plants to assure the necessary information flow for managing delivery priorities for customers.

For shop floor *scheduling*, the methodological criteria shown in Chap. 3 (machine/equipment loading evaluation) and Chap. 4 (workload assignment) are applied.

Referring to Vehicle Final Assembly Plants, in the following section, we will describe information technology systems applied for the management of production flows.

6.5 Logistics' Information Technology Systems

Methodologies described in the previous section require the utilization of modern information technology systems (ICT), with the application of software solutions, specifically developed and depending on company and process needs. These systems operate on two different levels:

- Central Level: process information necessary for operative planning in the medium term; data collection and elaboration is related to sales forecasts, productive capacity availability, management of industrial constraints, supply management and planning through MRP systems and inventory management of final products, in the plant and the market network. This information is normally processed on a monthly basis, consistent with OP rolling technique.

- Peripheral Level: process information necessary in each productive unit for managing working loads and production flows, and in each market area for managing delivery flows; data collection and elaboration is related to detailed planning of production (*scheduling),* daily and weekly line supply, daily and weekly dispatching of final and intermediate products (*dispatching)*, as well as planning for delivery of final product to the dealers network. For this purpose, systems are used for data collecting, processing and sending in real time or daily frequency.

The above information technology systems also allow for the automatic generation of a monthly report on the effectiveness of company logistic processes, according to parameters that will be described in Sect. 6.7.

Applying simulation models developed on the spot, it is possible to verify program feasibility quickly, searching for the best solutions for adjusting programs, according to temporal and industrial constraints. Proceeding this way, it is possible to speed up OP processing, according to the monthly rolling system.

Parameters to be considered in the development of software and hardware architecture are:

- complexity of product catalogues and correlation with *PDM* (products grid);
- complexity of final product in regard to subassemblies and modules (BOM).
- localization of plants and relative web network connection points;
- localization of delivery points and relative web network connection points;
- *lead times* for each of the logistic process phases that influence the lead time of order evasion;
- setting of production flow and controlling points, for each plant;
- criteria adopted for management of productive capacity constraints;
- criteria adopted for management of anticipations and delays in delivery;
- link to company economic and financial control.

Let us now examine some aspects related to information technology systems applied to the production process, with reference to body assembly, and paint and vehicle final assembly. The following chart shows, as an example, the main stations for flow setting and information diffusion to assure conformity of vehicle versions in relation to specific orders.

Flows related to space frames and body settings (operative pace and mix sequence) start from the insertion of main space frame subassemblies (station A), with *body framing* stations flexible enough for a mix of body versions to be produced.

Orders characterizing specific configurations of vehicles are associated with the bodies (and consequently with the serial number printed on the space frame), immediately before the bodies are introduced into the painting process, normally

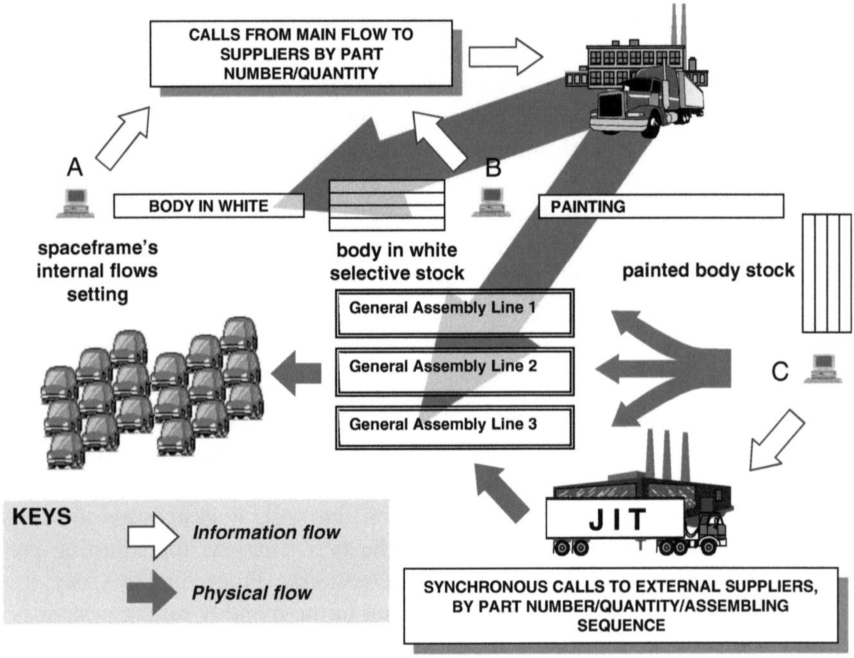

Fig. 6.8 Information technology system general assembly data setting and diffusion

organized by mini-batches for each color (station B), starting from selective *body in white* stock, sized on the spot and normally covering one hour of production Fig. 6.8.

Vehicle final assembly flow setting is led through selective stocks of painted bodies, also sized at a minimum level (station C), after having checked availability of main components (main mechanical groups and other components and subassemblies to be matched on the assembly line at different stages of assembly). Vehicles, during final assembly, have "specific cards", elaborated through the information technology system and specifying order requirements.

Finally, stations that determine flow setting and controlling are:

- space frame setting (A station)
- body framing
- body in white inventory
- external color setting (B station)
- painted bodies selective inventory
- diffusion to General Assembly for synchronous introduction of bodies and main mechanical groups to decking (C station)
- synchronous feeding of main body subassemblies to main assembly line
- end of line
- qualitative buy off, after functional test and sales line

- vehicle assigning and dispatching.

According to modern techniques, the above-mentioned systems for flow setting and controlling use:

- automatic devices for products and component "on process" identification (bar code systems, capacitive memory devices...);
- wireless systems for easy diffusion of information in real time to the different working stations for controlling and dispatching;
- monitoring networks interfaced with PLC and Numeric Control, which control machinery and equipment.

Similar systems are also used for engine, transmission and suspension assembly, because they present a high level of complexity, even if not directly connected to the final customer orders.

It is important to consider that the above IT systems allow for detecting and storing information related to supply batches and the manufacturing period, specifically for elements critical for safety *(Traceability)*. Every supplied batch and every temporal phase during the assembly process subject to traceability obligation is assigned to the relative space frame serial number on each vehicle or engine, so that it is possible to trace specific responsibilities in case of problems. Whenever it becomes necessary to intervene on commercialized vehicles, through recall campaigns, the traceability database allows for focalizing interventions at point of occurrence, minimizing the total number and their consequent costs as much as possible.

6.6 Key Performance Indicators of Logistics

As we saw in Chap. 3 regarding Manufacturing Systems, the appropriate key performance indicators (KPIs) of Logistics Systems can also be used to evaluate the effectiveness of processes and choose the right methods and actions in the technological layout setting phase for development of material handling.

For existing logistic systems, trend analysis of these parameters allows for evaluating effectiveness in orders, flows and inventory management.

6.6.1 Finished Product Inventory Indicators and Work In Process (WIP)

There are financial data and technical–economic parameters, related to working capital invested in the productive process. These indicators are typically:

1. **economic value of inventory (WIP)**, calculated applying the standard cost of production per unit to the quantity of elements in stock in the inventory and in progress along the process (according to manufacturing procedures);

2. **temporal value in working days (D)**, obtained by the ratio between the economic value of inventory and Work In Progress to the economic value of product developed per day, in the same productive process (production value at standard costs, delivered to warehouses for finished products or to customers);
3. **yearly rotation index (R)**, obtained by the ratio between the value of products delivered to customers in a year to the economic value of inventory and Work In Progress.

The trend analysis of the above indicators shows the logistic effectiveness from a financial point of view, according to the budget and improvement plans.

6.6.2 Process Lead Time and Flow Index (FI)

The Process Lead Time (PLT) corresponds to the process time between the moment of insertion of direct material into the productive process (most importantly, the withdrawal of raw or semi-finished materials from the warehouse) to the moment of the process's end. For assembly, the insertion moment is when the first component enters into the assembly process, while the end is when the finished products are delivered to dispatching points for customers or to warehouses for finished goods.

To evaluate the effectiveness of the production flow, the process lead time PLT can be evaluated statistically, distinguishing the main phases as follows:

- *type a* phases, during which the transformation process is active and operations adding value to the product are performed;
- *type b* phases, during which the product is transferred without adding any value;
- *type c* phases, during which the flow is interrupted, waiting for type a phases, because of repair interventions, process failures, or stocks and buffers existing inside the considered process (including inventory necessary for batch production).

Each of the above phases can be affected by the process's intrinsic variability, because of different standard times of operations depending on the product, variable waiting times for inventories, or eventual speed losses within the process.

Detecting these data statistically, the PLT average value can be determined and **flow effectiveness** also known as ***Flow Index (FI)*** can be measured:

$$PLT = \sum_{i=1}^{A} a_i + \sum_{j=1}^{B} b_j + \sum_{k=1}^{C} c_k$$

$$FI = \frac{\sum_{i=1}^{A} a_i}{PLT}$$

These calculations consider the total of A, B and C phases composing the analyzed productive flow, deployed into the above-mentioned typologies.

The analysis of parameter *FI* is necessary for searching for the best logistic organization of productive flow, choosing a more convenient method and layout. The more FI is close to one, the more the flow can be considered logistically lean, with reduction of PLT and speed-up of rotation index R of *work in process* internal to the productive process analyzed.

6.6.3 Risk Indicators of Inventory Obsolescence

These indicators are related to the following types of goods:

- spare parts for after sales technical assistance;
- spare parts for equipment and working means maintenance;
- other consumable materials used in industrial processes.

Through withdrawals and monitoring of materials in stock, the following can be automatically obtained:

1. **stock rotation index,** deployed by utilization classes and value segments according to ABC analysis criteria (see chapter on basic concepts at the end of the book);
2. **number of handled parts,** in stock and without request of utilization since "n" months (where "n" is the alarm threshold, according to the utilization mode);
3. **goods obsolescence index,** expressed as percentage of goods not handled (and probably no longer useful), for which it is necessary to calculate the devaluation and proceed to a partial or total disinvestment of inventories, having eliminated any possibility of re-utilization.

The trend analysis of indicators 1 and 2 can determine countermeasures for inventory optimization, assuring continuity of service and decreasing working capital, consequently avoiding risk of obsolescence.

6.6.4 Order Execution Lead Time (OELT)

We refer to a car's final assembly. The time necessary for the execution of orders starts from date of order acquisition (customer's obligation) and ends when the product is delivered to the same customer. We call the duration of this process Order Execution Lead Time (**OELT**), evaluated as a statistical data, referring to a specific model or product line, the final assembly plant and the commercial destination area.

OELT = Time necessary to introduce the order into the production plan, manufacture and deliver the product to the final customer

Let us see what the phases affecting the total duration of the execution of orders are:

(a) time necessary to introduce the order into final production process, considering available productive capacities in relation to order portfolio *(order processing time);*
(b) time necessary to end production process (*manufacturing lead time),* also including supply process of intermediate products affecting *lead time*;
(c) time necessary to transfer finished products from final assembly plants to commercial dealers, also including inventory waiting times necessary for organization of transport batches (*distribution lead time*);
(d) time necessary for sales operations at commercial dealers and public registration (*customer care time*).

To minimize Order Execution Lead Time, ideal conditions are:

- sales flow balanced with production flow;
- "one-piece-flow" production systems, with necessary degree of flexibility for multi-model production;
- production sites linked in an easy way to product dispatching centres;
- products diligently respecting order requirements, in terms of quality and specifications, without performing extraordinary finishing operations to correct manufacturing defects or damages occurring during transportation.

For mass produced cars, organized in a one-piece-flow, where ideal conditions are set, the order execution lead time is normally 3–4 weeks by best practices. In case of higher values, processes being slower and/or having a customer demand higher than production capacities, it is still important to define the necessary lead time precisely to maintain delivery obligations towards final customers.

As already established for *Time To Market* relative to new products, *OELT* for actual productions is also considered to be one of the strategic factors for sales success. When ideal conditions are not met, critical conditions can affect relationships between dealers and customers and there is a high risk of lost market shares.

6.6.5 Service Level (SL)

These indicators are strictly related to the satisfaction of obligations in terms of delivering on time and products corresponding to original requisites, according to order specifications.

Indicators normally used to evaluate Service Level are:

$(SL)_1$ **= number of deliveries on time/total number of deliveries (punctuality index)**

$(SL)_2$ **= number of corresponding deliveries/total number of deliveries (conformity index)**

Deviations from delivery obligations are evaluated through statistical criteria, dividing them into severity classes (1 or 2 weeks delay, 2–4 weeks delay, higher delays). These indicators can be referred to a specific industrial customer when we are dealing with materials or intermediate product supplies or to a specific commercial distribution area when we are dealing with final products, always in relation to the final assembly plant for a specific model or product line.

Indicator SL_1 is used to estimate punctuality of deliveries for component suppliers along the *Supply Chain*; this argument will be examined in Chap. 7, in the section on supplier performance evaluation.

Let us conclude this section underlining that the main *Logistic Key Performance Indicators* are those relative to Finished Products and WIP inventories, to Service Level towards commercial networks and Order Execution Lead Time. Modern information technology systems allow for automatic data collection, assuring a monthly report.

Other indicators are not subject to recurrent data collection and reporting, but are specifically analysed in process phases and the setting up of manufacturing systems or in considering specific improvement programs for logistic processes.

6.7 Just in Time Approach

As mentioned in the historical outlines in the first section of this chapter, starting in the '70s, Japanese industries, predominantly Toyota, developed a peculiar approach to logistics called Just In Time (JIT); this method of production was a consequence of specific constraints affecting the automotive industry in Japan:

- *small spaces* available in plants;
- *product complexity*, increasing of model mix over the years;
- *worldwide markets* scenario, with several competitors;
- *cost reduction* needed to increase profit margin;
- *service level* towards customers to be increased (lead time to be reduced at all levels).

Starting from these items, Taiichi Ohno, at that time responsible for manufacturing for Toyota Motors, took some basic concepts of massive distribution from observing the American industry's big supermarket approach to business and created the JIT approach.

This approach is based on the principle of producing only what customers require, at the time it is required, under the conditions required and in the quantity required. To apply this concept, it is necessary first to have a continuous flow in the internal process and then search for the synchronization of internal processes (at all levels) with customer demand, by transmitting information at the shop floor level from customers to suppliers through the whole manufacturing process, which should always be regular without failures (machines breakdowns, human errors).

These basic rules can be seen in the following **JIT four principles**:

1. **Continuous One Piece Flow**: process should be set to have one single piece moving from one stage of the process to the next without breaking the sequence and without waiting times along the way.
2. **Takt Time**: the rhythm of the process should be set considering customer demand (orders if possible) during a certain period of time and leveling production (*heijunka*) by playing on shifting peaks of customer requirements when downtimes in production are visible to create a constant production request for the longest time possible; pace of production can be calculated by the ratio between net available working time in the period considered and customer demand.
3. **Pull**: production must be performed by pulling the products requested by customers from the end of the assembly process up to the incoming raw and semi-finished materials by moving the information process to process. For this purpose, Ohno introduced the kanban (little tag) system, which carries information of the requested production to the previous step of the process, applying the principle used in American supermarkets to replace only those goods withdrawn from shelves directly by customers.
4. **Zero Errors**: this means having no defects on the products due to human errors or equipment failures; if an error occurs during production, the process must be stopped to avoid the production of a defective product that will have to be reworked once out of the process; production can be restarted only if conditions are fixed.

These concepts, seen in the logic of JIT, have become one of the main pillars of the Toyota Production System and, at the end of the '80s, also the basis of the Lean Manufacturing logic applied worldwide beginning in the USA.

The international bibliography on these topics is very rich, and readers can find more details in multiple publications by Japanese, American and European writers.

Chapter 7
Global Purchasing Operations

7.1 Role of the Purchasing Department

As seen in Chap. 1, industrial policies adopted by carmakers during recent decades have resulted in a higher involvement of suppliers in the development of product and production processes. Referring to the world's foremost carmakers, we note that their *make or buy* policies have been consequential for the following items:

- in engine production, according to schemes discussed in Sect. 1.5, the shares for "buy" stay within 65 % and 70 % of the total product cost;
- in final assembly of automobiles, according to schemes discussed in Sect. 1.5, the shares for "buy" stay within 75 % and 80 % of the total product cost (excluding engines from the calculation).

The supplier's obligation then becomes determinant for reaching final product competitiveness. For this purpose, we dedicate a whole chapter to the topic of purchasing and industrial cooperation, dealing with the methodological criteria for purchasing applied by carmakers rather than with the capacity of component and working means producers to correspond to these criteria.

In major industries, particular automotive compartments, the process of making purchasing decisions is shared by a number of departments within the company, with the assistance of a cross-functional committee or project groups (product platforms).

The Purchasing Department is in charge of managing the ordering process, from choice of suppliers to gauging satisfaction in regard to cost, quality and punctuality from same. The decision-making process, even if shared with other departments, must be coherent with Top Management Purchasing policies, assuring a certain ethic in relations with suppliers to protect company business interests.

In big companies, both for reasons of function and need for control, ordering and contract obligations respecting verification processes involve three subjects:

M. Gobetto, *Operations Management in Automotive Industries*,

Springer Series in Advanced Manufacturing, DOI: 10.1007/978-94-007-7593-0_7,

- *Technical Purchasing Department* (authorized to spend), in charge of purchasing orders, according to necessary technical and economic information, in relation to the target of industrial initiatives of competence;
- *Purchasing Department,* in charge of procuring agreements with suppliers, in relation to the production plan, assuring best market conditions;
- *User Department* (production or technical systems manager), in charge of authorizing bill payment after delivery of goods or services has been provided, corresponding to orders made by the Purchasing Department and to the supply plan.

The Purchasing Department must also take part in management of contract relations, solving eventual problems, so as to avoid, if possible, legal action between parties.

Let us remember that an enterprise's processes involve several types of material and industrial resources; these are outsourced and can be classified as follows:

- *direct materials*, subject to transformation and assembly operations during the production process and composition of the final product; direct materials are purchased by quantity when they are "finished elements" by drawing specifications, or by "weight or volume" when they are "raw or undefined materials", even if determined by precise technical standards;
- *energetic vectors and consumable materials*, not composing final product, but used for industrial processes;
- *services*, used for administrative and industrial processes (engineering, logistics, maintenance, after sales technical assistance, information technology services…);
- *other goods included in fixed capital,* according to classifications laid out in Sect. 6.1.

In this field, the Purchasing Department is in charge of the following fundamental tasks:

1) *selecting and strengthening set of suppliers*, applying evaluation criteria shared with User Departments;
2) *procuring offers, coordinating technical–economic comparisons and proceeding to purchasing tenders*;
3) *leading negotiations and making orders*, based on technical specifications and targets set in specific initiatives;
4) *managing contractual relations with suppliers.*

To resume, the Purchasing Department is the "linking point" between the Company Purchase Departments (plants, technical offices, project leader…) and suppliers for materials, services and industrial goods. The relative contractual obligations are fine-tuned by the Purchasing Department, by proxy for Top Management or Business Unit. For management of contractual relationships, the Purchasing Department is assisted by certain staff departments, such as Finance and Administration, Legal Services, and Quality Assurance.

As seen in Chap. 6, logistic effectiveness is particularly relevant in supply policy. In fact, supply and the reduction of technical lead times along the supply chain is determinant for sales success and for reduction of work in progress.

Finally, the Purchasing Department must procure agreements from suppliers to obtain the following results:

- Secure sources/adequate stocks for availability of direct materials and consumables
- New product development planning in respect to production means and availability of systems
- Flexible production systems, "just in time"sourcing methodologies for effective logistic process for management of production planning
- Optimal localization of plants and fast flow for effective delivery of goods

The Purchasing Department is also in charge of preventing critical situations, such as those derived from a supplier's financial problems, by using necessary countermeasures to avoid sourcing interruptions. For this purpose, other Departments are in charge of supporting this task (Finance and Administration, Legal Services and Corporate Business Management Staff Department).

7.2 Evolution of Purchasing Policies

It is right to link this topic to what we discussed in Chap. 1 regarding "make-or-buy" policies and technological aspects developed autonomously by Specialized Suppliers (Sect. 1.5).

It is to be underlined that the actual trend in vehicle development, as much as possible in a modular concept, on one hand simplifies final assembly operations and on the other obliges first level suppliers to localize their productive activities near the Final Assembly Plants.

Purchasing policies adopted recently by "carmakers" tend to create stable and even more engaging relationships, according to the "extended enterprise" model, which involves first level suppliers in "co-makership". As a consequence, less expensive and higher quality supplies are obtained through open agreements over the long term and through shared efficiency plans, in relation to final product competitiveness targets.

Changes that occurred in the main "carmaking" organizations and new trends affirming the supply of vehicle space frame modules (brakes-suspension modules) are summarized in the scheme in Fig. 7.1.

Another example is represented by vehicle air conditioning systems: from the procurement of single components to complete conditioning subsystems, made by components partially connected to cockpit modules (controls, controlling devices and secondary thermal traders), partially to the vehicle's "front-end"(sensors and primary thermal traders) and partially to the engine (compressor). Pipes for

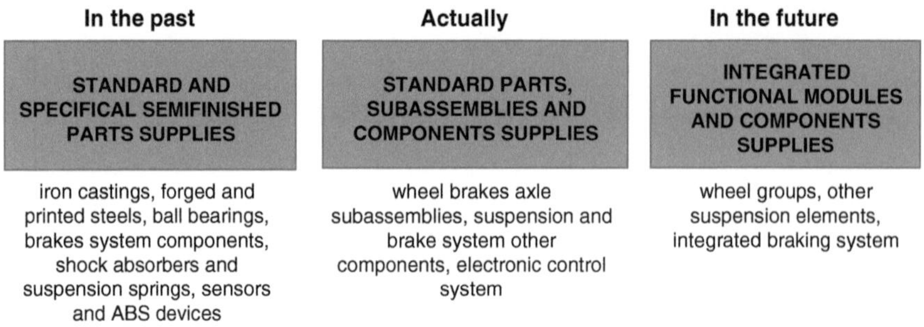

Fig. 7.1 Carmakers historical trend for global sourcing

air diffusion and relative regulation systems inside the vehicle are developed in "co-design".

First level suppliers are even more involved as co-makers, sharing new models launched by the "carmakers" risks and opportunities. At the right time, component producers must try to expand their deliveries to more "carmakers" so that they can reach a good economy of scale, avoiding dependence on only "one customer".

"Co-design" is a critical factor for success in "co-makership" and can be achieved by extending Simultaneous Engineering methodologies to First Level Suppliers, according to methods indicated in Chap. 2. **A First Level Supplier receives from the Purchaser prescriptions for functionality of components and subassemblies to be developed, and proceeds autonomously in industrialization, up to the final supplied product qualification, engaging to respect development programs, obtaining established functionality and quality requirements (Product Functional Deployment and Quality Rate, according to what was discussed in** Sect. 2.1) **and assuring availability of quantities at the shared cost.**

In Table 7.1, know-how profiles are laid out for cooperation between carmakers and suppliers, related to the development of the vehicle's components and subassemblies.

Modules and subsystem suppliers must have an appropriate engineering and research centre; they must be able to coordinate their sub-suppliers, assuring necessary integration; to determine product functions, they must also set a technical after sales network, to support the carmaker properly in the several geographical areas of vehicle commercialization.

In choosing suppliers in class A), B) and D), the capacity to promote innovative solutions and supply competitive products is the first consideration; another important point to evaluate is the availability of productive sites near the carmaker's Final Assembly Plants (whenever logistic complexities and special customers' taxes are in place).

In Table 7.2, purchasing policies actually used by major carmakers in relation to the respective targets are shown.

Table 7.1 Carmakers—suppliers co-operations

A	Modules and subsystems supplier, co-design with "carmaker"	• More capacity for technological integration • Autonomous capacity for development and experimentation • Establishment of project's attitude, with setting of challenge targets • Market leadership in business area
B	Components supplier, developed autonomously and in relation to specific applications	• High technological specialization • Autonomy in development of standard solutions • Wide customer range • High economy of scale
C	Components supplier with carmaker design	• Design contribution focused on manufacturability • Ownership of production processes • Lean production organization
D	Standard components and raw materials supplier	• Capacity to offer materials with current technologies • Capacity to offer materials with continuity and at competitive prices, according to market trend

Table 7.2 Carmakers purchasing policies

Policies	Objectives
Rigorous selection of suppliers based on innovation capacity, qualitative levels and competitiveness of products and services offered	Concentration on a few highly qualified suppliers, with advantage of economy of scale and avoiding monopoly positions Supply management simplification
Development of range of suppliers on a worldwide basis, switching to dedicated and, when possible, localized industrial sites	Activate supplies from worldwide industrial areas, based on lower industrial cost logic Following vehicles' productive capacity development in the strategic worldwide markets
Involving Suppliers for product innovation and to realize highly competitive productive capacities	Offer final product consistent with market requirements to customers Shorten new products' "time-to-market" Reduce carmaker investments
Use of modern WEB techniques and "e-Procurement"methodologies	Make easier exchanges on a worldwide basis Optimize purchasing tenders Minimize structures dedicated to supplies

The matrix shown in Fig. 7.2 **lays out the logic for suppliers' choice and industrial co-operations and the development decision-making process.** Please note that while Supplier Know-How reflects the level of technological complexity of the supplies, the Supply Logistics' Complexity reflects exactly what was examined in Sect. 6.3 in terms of material handling (material classification).

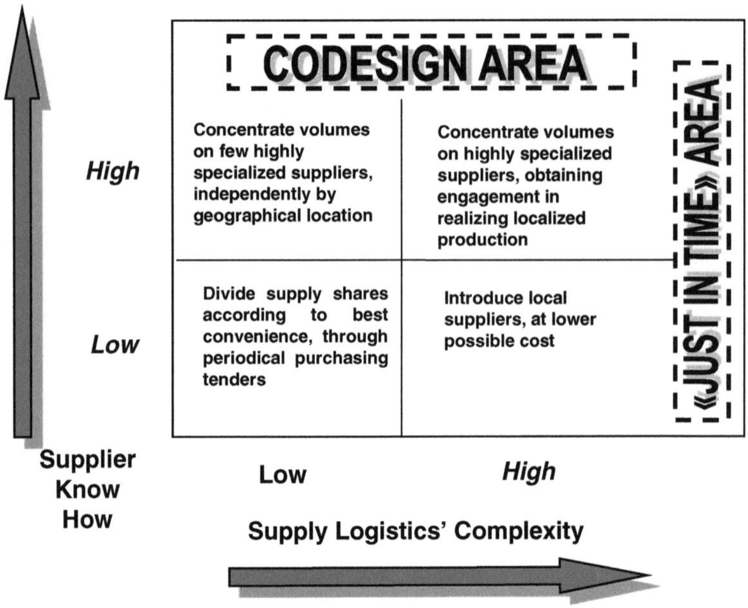

Fig. 7.2 Carmakers suppliers' selection logic

7.3 Purchasing, Marketing and Global Sourcing Policies

Market conditions for purchasing are obviously influenced by the offer/demand ratio, referring to the specific technological compartment.

The possibilities for purchasers to obtain the best supply conditions increase when there are more alternative solutions and product offers comparable to market demand. Conversely, the Purchaser is disadvantaged when valid and ready alternatives are missing, or when product demand is higher than market offer.

In specific technological areas, characterized by high specialization and innovation, it is possible to have exclusive supply conditions, concentrated on a few suppliers that operate as market controllers. In such situations, "new entry" producers can be pushed by supply synergies shared and adopted by the major carmakers.

Considering the above demand/offer ratio, for each type of product and technological specialty, marketing analysis must focus on:

- actual foreseen dimension for demand;
- actual foreseen dimension for offer;
- competitors' picture and respective market positions: "leader", "follower", "new-entry"...
- the degree of independence of alternative suppliers;

- the availability of suppliers chosen for long and stable relationships through long term agreements;
- availability to actuate investments consistent with activity levels required;
- risks derived from eventual "single sourcing".

Another important topic of purchasing strategy is the globalization of international markets, which is the "global sourcing" policy.

Through studying established multi-national enterprises, the primary reasons for widening supply sources at a worldwide level can be detailed as follows:

- **reducing the incidence of developing costs;**
- **encouraging localization of productive capacities in areas with lower labour costs, for "labour intensive" productions;**
- **capitalizing/sharing "best practices" between geographical poles of supply;**
- **localizing logistically critical component production near final assembly plants;**
- **obtaining the best economic conditions for supply on a worldwide basis.**

Furthermore, it is necessary to improve the qualitative level of supplies towards the best and most homogeneous standards, by operating as follows:

- levelling evaluation criteria and procedures of suppliers;
- leading international development of available suppliers;
- promoting the transference of "know-how" to suppliers in new countries.

Analysis of purchasing markets begins with knowledge of the following macro-economic and financial aspects, characterizing geo-political areas in which new supply sources have been activated:

- industrial growth trends;
- interest in new local investment;
- exchanges and monetary stability;
- availability of funding, with competitive conditions;
- inter-governmental commercial agreements and customer constraints.

The decision to widen supply sources on a worldwide basis is a consequence of the following strategic evaluations:

1. **availability of raw materials and energy and relative supply costs;**
2. **availability of a trained workforce at low cost;**
3. **ways, time and costs for moving final products to points of use;**
4. **tax concessions for new investments;**
5. **social and political context favourable to industrial development.**

These evaluations regarding macro-economic, social and environmental aspects do not only refer to existing conditions, but also to a medium-long term trend.

7.4 Management of Suppliers Network

A suppliers network is made up of enterprises that co-operate with the Final Producer, according to the "supply chain" structure described in Chap. 6.

Each Industrial Group producing vehicles, components, and equipment has its own "suppliers network", made up of enterprises charged with supplying direct materials, consumables, working means and services. The availability of a "suppliers network" suitable to the enterprise's activity is a very important share of the "goodwill", considering the intrinsic start-up value and the contribution that suppliers can give to the development of new products/processes.

Information support is made by a database of companies qualified to supply in specific areas of competence (Vendor List). This database is divided per product and/or technological area, with reference to what was discussed in Sect. 1.5. In each merchandise area, the qualifications of suppliers are assessed, based on the evaluation process shown below. Eventual critical suppliers are assessed as well; these are considered under observation, authorizing them for actual supplies but excluding them from new business until they have solved their criticalities.

Evaluation Process

Both "Carmakers" and their suppliers of components and working systems have developed precise certification procedures, based on the application of shared international norms.

The point of reference is the international norms ISO 9000 that oblige enterprises to adopt homogeneous criteria for evaluation of quality levels.

Nevertheless, ISO standards, being general, are not precise enough to define a supplier's qualification process in an automotive compartment. As an example, we mention here the following norms: ISO 16494 and QS 9000, both developed by the "big three" American companies (General Motors, Ford and Chrysler).

The supplier evaluation process deals with the following aspects:

1. **supplied products and quality service level;**
2. **reliability of new product development plans;**
3. **service levels related to continuity and conformity of supplies, according to plans.**

Relative to the *first aspect*, for direct material supply, "carmakers" use the following indicators:

- rejects/defects index in parts per million, statistically measured within homogeneous material or product families;
- incidence of costs for "non quality" of defective materials or due to technical inefficiencies of products or services supplied;
- costs for "non quality" due to recovering campaign or to severe repair interventions in after sales, when supplier's responsibility has been verified.

This last factor is considered one of the most important and can determine severe actions towards the supplier, especially if brand image can be affected.

For each one of the above statistical data, monthly and yearly reports are done, comparing suppliers and applying homogeneous evaluation criteria.

Relative to the *second aspect*, we have to emphasize that the evaluation deals with the "key points"of the industrialization process (see Sect. 2.2). For example, delays in development of a new manufacturing system can be determined by a delay in production start-up and in commercial launch, with loss of profit margin and damage to the brand image.

Relative to the *third aspect*, criteria shown for logistics in Chap. 6 apply.

A supplier evaluation is done in advance to supply decisions and is a determinant element for the assigning of initiatives. Then, it must be found in objective and transparent data in regard to the enterprises that are deemed unreliable.

Within the "Suppliers Network", the choice of which suppliers will be more engaged and involved in the development and launch of a new product is charged to the Purchasing Department in agreement with the Company User Department, based on references and evaluations made through the following requirements:

a) **capacity to correspond to product targets (Product Function Deployment) and to guarantee stable quality levels;**
b) **competitive prices, compared to best market conditions and within specific benchmarks or tenders;**
c) **availability of corresponding necessary investments to guarantee required productive capacities.**

The basic element for evaluating the Purchaser/Supplier relationship is the achievement of the new product's targets (cost, quality, and development lead times). The evaluation of "performances" deals with key points of respect of the industrialization process (seen in Sect. 2.2) and the regular deliveries established through agreed programmes.

It is useful for the Top Management of major carmakers to establish "excellence awards" for suppliers who have shown they are capable of good results and a good level of service. These are very sought after prizes, because they represent an important reference on the market.

For the introduction of a new supplier, after marketing analysis and based on tenders, through the collection of references and technical assessments, it is important to evaluate:

- financial solidity and structural suitability of a company;
- availability of R&D structures, whenever there are co-operations in future technical projects;
- availability of technologies and manufacturing systems, specifically related to supplies;
- quality system qualifications, through an external entity allowed to verify international certifications (ISO 16494, QS 9000…).

After these evaluations, the introduction of the new supplier is delivered based on economic advantages obtainable from actual production and/or the capacity to develop new products competitively.

7.5 Order Procedure and Cooperation Agreements

The contract relationship between a Purchaser and a Supplier is based on emitted orders and the supply conditions specified in it (generally technical specifications, with reference to drawings and control methods attached). Normally, general conditions, applied by the Purchaser to all suppliers operating in the same compartment of activity, are also included. The offer and negotiation process, preliminary to order emission, can lead to the modification of the initial requirement of the Purchaser, if judged inconvenient for both parts.

Ordering procedures can be distinguished as:

1. **Closed orders**: applied for purchasing when there is a specific order for which negotiation is periodically employed, or when products are ordered from "a catalogue", according to a price list. The most relevant part of closed orders is related to working tools and equipment.
2. **Open Orders**: applied for direct and consumable materials, executed through material requirement planning procedures, according to MRP (see Chap. 6). For this purpose, specific order cards are set, shared with the Purchasing Department and uploaded into the information technology system; these cards show the shares assigned to each supplier for each product code. Prices must be defined by the Purchasing Department, with a confidential level on the information technology system and administrated by the Administration and Finance Departments.

Relative to direct materials and part components, a supplier's contract obligations consist in guaranteeing the Purchaser of:

- **required production capacity, according to agreed programmes;**
- **established quality levels, based on drawings and technical specifications;**
- **agreed-upon supply and delivery conditions;**
- **respect of new product development plans, according to shared targets;**
- **prices maintained for the entire agreed-upon period (with eventual issues of variation linked to labour and raw material costs, according to the recognised statistical index);**
- **efficiency and cost reduction programmes, agreed relative to the contract period.**

At the same time, obligations charged to the Purchaser consist mostly of:

- **assigning shares of production planning as agreed in the contract, for the length of the agreements;**

- **withdrawing delivered products (if responding to agreed-upon quality levels and requirement planning transmitted) and consequently paying under established conditions;**
- **recognizing activities for product/process development (if established target has been achieved) and consequently paying under established conditions.**

Other contractual obligations can be related to cooperation for Research and Development for Product Improvement or recognition of patent or licence, or other specific warranty conditions.

Expenses necessary for R&D activities charged to suppliers and for the specific tooling of new products are partially or totally to be sustained by the Purchaser, together with negotiations over the purchase of components.

The relative remuneration (una tantum) is managed by the Purchaser with the following payment alternatives:

1. by eventual anticipation of each step of activity improvement and final payment at delivery of product, considering product qualification and fulfilment of final check;
2. by "shares" on quantities of products delivered progressively and up to the extinguishment of agreed payment;
3. by other method of delayed payment, during the product life cycle (normally through established annual shares).

When the Purchaser is in charge of specific tools, they are included in the equipment database (invested capital), even if they are rented to the Supplier, who is in charge of providing for necessary maintenance to assure normal technical life.

We underline that payment methods are normally a financial leverage for the Purchaser, considering the lead time between delivery of the goods and accounting of the relative remuneration.

Order emission starts the information flow for supply management and for the accounting charged to the Purchaser. Modern information technology systems allow for automatically matching the receiving note (delivery document) to the invoice made by the Supplier, to be considered as anticipation or expiration of the same order, in case of "closed orders". If this information technology system is reliable and protected from external risks, the accounting process is made automatically, simplifying the invoice verification process.

7.6 Supplying Cost and Purchasing Effectiveness Indicators

At this point, we deal with the topics of "purchasing price" and "supply cost," the latter related to direct and consumable materials.

Purchasing price is only a part of total supply cost, even if it is the most important part.

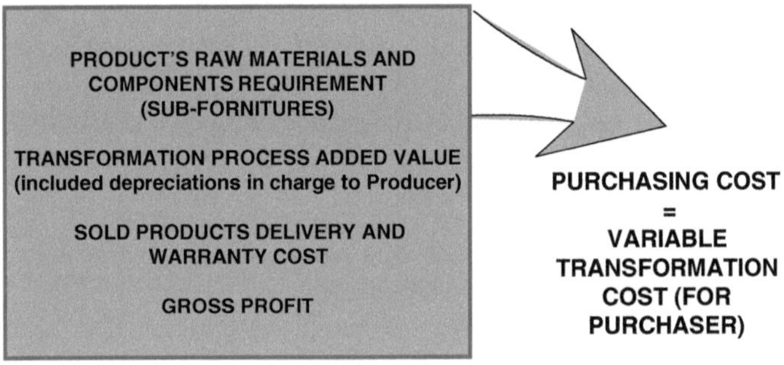

Fig. 7.3 Cost of sale composition for a supplier

A lower purchasing price does not necessarily lead to a lower supply cost, in case of higher costs due to specific delivery conditions or supplied product quality level (production rejects or reworking during production process or in after sales assistance). In critical situations, it is necessary to investigate on "non quality"and/or "low service level" related costs.

The scheme in Fig. 7.3 represents the purchasing cost composition that is normally the most relevant part of User/Purchaser "variable transformation cost".

The gross profit is to cover the company's general expenses charged to the supplier, including R&D (except what is charged to the purchaser, on specific initiative), as well as taxes and working capital remuneration.

During negotiations, the "buyer" tries to obtain from the supplier a detailed proposal about costs, so that benchmarking analysis can be easily run. This is necessary specifically when it is not possible to use purchasing tenders, as a consequence of constraints on supply assignment. Either way, the "buyer" normally operates in relation to well-defined cost targets, assigned by the Purchaser Department (once cost/benefit analysis has been run).

Let us now see the composition of total supply cost for "direct materials", as shown in Table 7.3

To evaluate the supply process of direct materials, it is necessary to compare the several supply sources, considering all the above items of cost.

Normally, an accounting system is unable to process data related to indirect costs automatically (bullets 2., 3., 4. and 5.). As a result, it is necessary to run specific extra-accounting analysis, to examine details of specific critical situations.

Indicators normally used to evaluate Purchasing Department "performance" are:

A. *Economic Indicators*

1) **Purchasing Effectiveness** = annual incidence of economy on purchasing price, considering the same product and raw materials and labour cost factor trends, according to statistical data recognized by the company.
2) **Cost Target Achievement** relative to new products (evaluation shared with the company's other departments in charge of the Purchasing process).

Table 7.3 Composition of total supply cost

1. Supply direct costs	Purchasing price, included cost for packaging and transport for deliveries to the facilities charged to the suppliers
2. Indirect costs for logistics management	Costs for transport to delivery points near to final user plants, costs for receiving, warehousing and transferring of material to manufacturing systems
3. Costs for "non quality"	Costs consequent to defective materials and components of supplied products
4. Costs for "lack of service"	Costs consequent to interruptions in production process due to delays in delivery
5. Fixed costs for "material management" departments	Purchasing, direct material management and arrival material qualification department

The above indicators are checked and elaborated by the Finance and Administration Departments.

B. *Indirect Indicators*

3) **Supply Quality Level Index** (comparison with annual average values) according to statistical evaluations made by the Quality Department.
4) **Supplier Service Level Index**, according to statistical evaluations made by the Supply Chain Management Department, by applying criteria demonstrated in Sect. 6.6.

C. *Functional Indicators*

5) **Lead Time** between purchasing request and order emission "on initiative".
6) **Supplier Range Management Effectiveness**, equal to the reduction of number of suppliers and the "turn-over", obtained by selecting and introducing new competitive sources of supply.
7) **Purchasing Burden Cost Incidence** on value of managed supplies.

In addition to the above-mentioned indicators, it is important to estimate the achievement of quality targets and new product "time-to-market", which depends on cooperation of the supplier.

7.7 "E-procurement" Techniques

In Sect. 2.3, we saw the importance of an information technology system for integrated production system management (PDM system). The order acquisition process is also included in this.

The introduction of "E-Business" techniques, supported by the Internet and Intranet, allow for an efficient process of research and assignment of orders to the suppliers, particularly for direct and consumable materials, the latter treated as

Commodities, and for standard working means. For this reason, we deal with utilization of these techniques in the following ways.

By "e-Procurement", we mean the order acquisition process for goods and services using WEB networks and modern "information technologies systems" described in Chap. 2: the process begins with research into the most convenient supplier, including order acquisition and emission (consequent to transaction), and ending with final delivery verification, which activates automatic payment and invoicing.

A Selected Suppliers catalogue is managed through the company's intranet, so that transactions and orders can be done directly, assuring privacy of information and constant control of the supply process through the company's information technology system.

Benefits obtainable from the utilization of a modern "e-Procurement" system can be:

- optimization of processes, with purchasing cost and time reduction;
- "on line" availability of updated information about goods and service offers;
- possibility of running purchasing tenders through "on-line pools";
- automation of communication between Suppliers and Users, even those at great distances;
- electronic management of documents (offer requirements, orders, technical specifications…), reducing the need for paper documentation;
- integration between Purchaser and Supplier information technology systems.

This method allows minimization of cost related to information and administration in Purchaser/Supplier relationships. The difficulties in the adoption of "e-Procurement" are due to the characteristics of pre-existing information technology systems and data protection issues.

The utilization of these techniques is very useful for the acquisition process but cannot be independent of a supplier's evaluation criteria for capacity and availability, as specified in Sect. 7.4.

Chapter 8
Quality Management and Continuous Improvement

8.1 The "Learning Curve" Concept

Innovations and new technologies applied to new products improve by increasing activity levels and applications, as well as applying criteria for industrial improvement.

Historically, the concept of the learning curve was tested for the first time in plane factories during the Second World War, when it was noticed that working times and related costs were decreasing in a regular and progressive trend if intensive productive plans were applied, according to a predictable trend on a statistical basis. From that moment on, the learning curve concept has been considered valid for several industrial applications, especially in conjunction with a high level of technology.

This concept is connected to "standard logic", as seen in Chap. 2; in fact, once the testing and research phase is complete, for every new product, industries refine technologies and manufacturing methodologies, searching for the most convenient, allowing for higher economy of scale as the product is affirmed on the market.

In general terms, it is possible to trace the "learning curve" for a new product, predicting it through comparison with that of a similar previous product. In this way, it is possible to evaluate the cost reduction over time in relation to the progressively produced quantity. The unitary cost reduction is expressed in percentage for every time the progressively produced quantity is doubled. Practically, a "learning curve" with a negative apex sign means that the unitary cost decreases 10 % at every doubling of the cumulated productive volume.

Thus, the arithmetical formula that describes this trend is an exponential one:

$$\mathbf{y} = \mathbf{a} + \mathbf{b} \cdot \mathbf{x}^{(-\mathbf{k})}, \quad \text{where:}$$

a is the final unitary cost when the product will be fully affirmed on the market;
b is the additional unitary cost recorded in the first production batch;
x is the progressive number of homogeneous batches;

M. Gobetto, *Operations Management in Automotive Industries*, Springer Series in Advanced Manufacturing, DOI: 10.1007/978-94-007-7593-0_8,

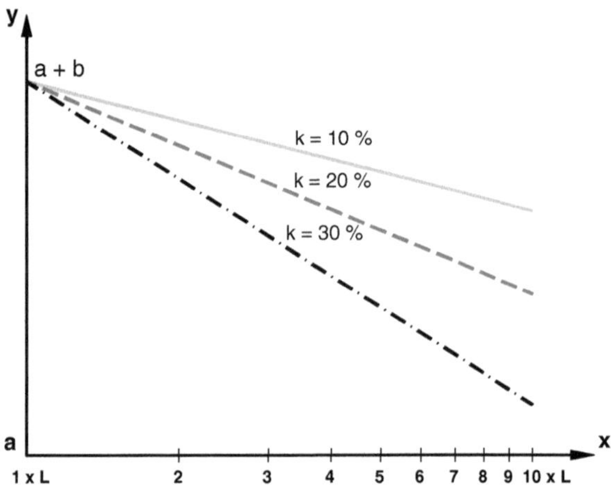

Fig. 8.1 Learning curves example

y is the unitary cost related to the progressive batch x;
k is the exponential apex that determines the decrease of the unitary cost in relation to the progressive production quantity developed.

The exponential apex (k) is called the *cost decrease index,* as a consequence of the "learning curve."

By using a logarithmic scale for the product unitary cost in relation to the progressively produced quantity, the diagram represented in Fig. 8.1 shows decreasing linear straight lines, in which k represents the slope, as in the next chart.

We emphasize that the *cost decrease index* (*k*) is higher when applied technologies are more innovative.

As the above formula is derived from a statistical analysis, the "learning curve" also represents the capacity to improve a specific product progressively, from the start-up phase to its "industrial maturity."

This trend is related not only to the standard manufacturing cost, but also to quality parameters (production rejects incidence, defect rate during product life, duration reliability).

8.2 Production Ramp up and Maximum Rate Achievement

The so-called industrial logic of "continuous improvement", which we will deal with in this chapter, is based on the above-mentioned "learning curve" concept.

Relative to the Manufacturing Systems, the "learning curve" is mainly affected by the speed with which the workers learn and by working time needed for technical settlement. For mass production, in tracing the "learning curve" through prediction, we assume the available productive capacity (APC, according to the definition in Sect. 3.2) as a reference.

The diagram on the following page shows the real working times trend (ST/ŋ), in relation to the progressive productive volume. Based on historical statistical data (selected by rejecting anomalous cases), we assume that:

- the "learning curve" will follow a specific coefficient k (for example: 20 % for each doubling of the progressive productive volume);
- total time for technical and organizational settlement needed will be x days of production at maximum rate (in the example, x corresponds to 50 times the planned daily productive capacity);
- labour and working means requirements established by a manufacturing engineering plan will be quickly available.

The diagram in Fig. 8.2 shows the standard working time trend (y = ST/ŋ), in relation to progressive volume produced (x).

x = Volume progressively produced, expressed in days of production, obtainable at maximum production rate (APD).

y = Real working time employed for product unit (ST/η).

y = 1 is the standard value at maximum rate, assuming that the product and the productive system will be set to the progressive value 50 × (APC), being APC the Available Productive Capacity per day.

In our case, considering a vehicle's final assembly line (labour intensive), the main factors influencing the "learning curve" are:

- **degree of product complexity** (for example: the final assembly of a top range car, with a high level of content, requires a more demanding "learning curve" compared to that of a lower level car);

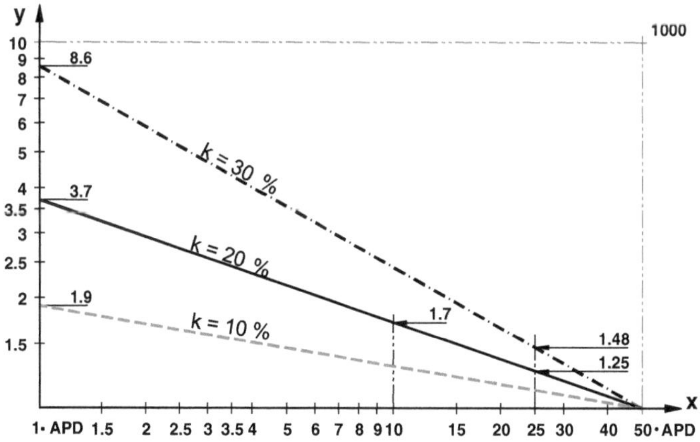

Fig. 8.2 Standard working time trend

- **worker attitude and speed in learning**, for which the workers are progressively able to avoid errors and display expertise in assembly operations complying with the foreseen standard times (ST);
- **manager coaching and stretching ability.**

Considering the "learning curve" to be a characteristic element of the specific product and process, we can trace the "production ramp-up curve", as shown in Fig. 8.3.

Phase (a) **Product/process technical try out** It starts from "job one"—i.e., from the first technically acceptable production—and ends with the achievement of "process capability" targets. In this phase, some components can still miss quality targets and some equipment can have failures. To avoid negative impacts on products delivered to the customers, specific quality checks and functional tests are performed.

Phase (b) **Fast ramp up** The curve slope is influenced by the possibility of feeding components for assembly with regularity and introducing more workers, initiating "on the job learning" activities through the use of expert workers.

Phase (c) **Technical and organizational optimization** The duration of this phase depends on the difficulty in achieving the standard working times foreseen by the project and obtaining the full effectiveness of the working force. Normally, the last 25 % of the production ramp up, considering the same workforce, can be achieved according to the "learning curve".

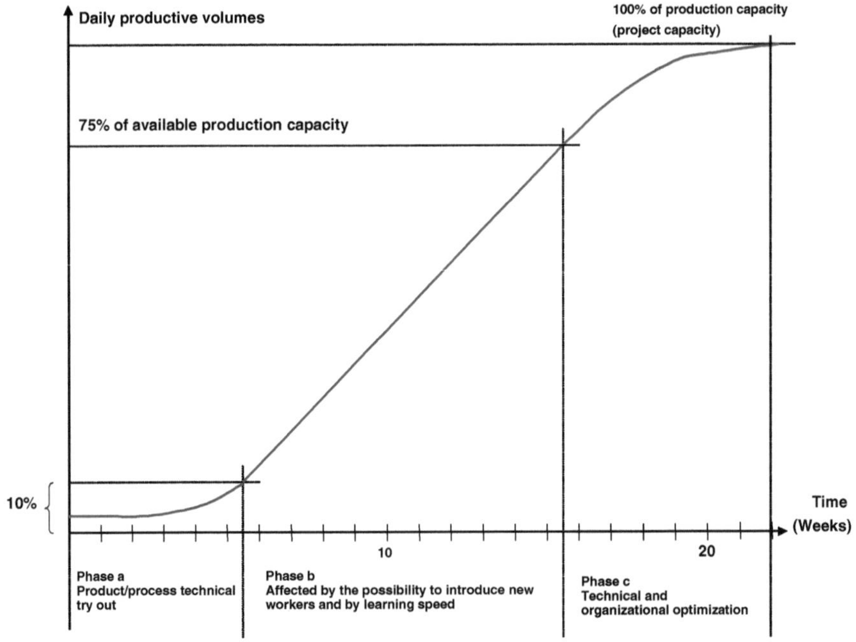

Fig. 8.3 General assembly production ramp-up curve example

For production operations that concern complex, capital intensive processes, the production ramp up speed and time necessary to achieve the maximum rate foreseen by the project depends mainly on the reliability and maintainability of the manufacturing systems, according to what was discussed in Sect. 5.2.

The production ramp up speed determines the time needed for start-up of series production (job one) and the date of achievement of the maximum planned production rate. Let us remember that this parameter heavily influences the new product's "time-to-market."

8.3 Quality Management in Industrial Processes

Quality for a product or service is measured directly by Customer Satisfaction, in much the same way that reliability can be described as the capacity of the product to perform its function over time in a continuous way, according to the customer's needs, assuming that utilization and maintenance rules, defined by the product maker, will be respected.

Product Quality at Origin and *Quality of Service in After Sales* are determinant for achieving the customer's trust and maintaining fidelity for future ordinations.

This is true not only for the Producer/Final Customer relationship, but also between all enterprises along the "supply chain" up to the dealers' network, according to the concept of "extended enterprise."

Even inside the same enterprise, each department or productive unit must try not to create problems for their internal customers, both in supplying sub-assemblies, working means or services and in transmitting technical or managerial information. This philosophy inspires *Total Quality Management* (which we will discuss in the following section), to which the behaviour of all involved should correspond.

In this section, we will deal with the improvement of Industrial Process Quality, assuming that a normative ISO for Quality has been achieved.

We will also consider the known concept of "Process Capability" and the meaning of the parameters Cp (determined by intrinsic dispersion, according to the Gauss curve) and Cpk (determined by the positioning of dispersions in relation to the limits of the range of tolerance admitted). Let us see which is the best way to obtain Cp and Cpk, as defined by the project and in relation to the components and final product quality targets:

1. During project design, the "manufacturing engineering plan" is set, in relation to the product's technical specifications, by selecting the best technical solutions for assuring tolerances defined by drawings. For this purpose, it is necessary to verify the possibility of obtaining these restrictions in advance, according to manufacturing "Process Capability".

As a result of this verification, manufacturing methodologies are defined and part design can be refined by competing technical specifications necessary for developing manufacturing systems.
For example, to contain the flatness and parallelism errors of brake disk friction, the operation of grinding on braking bends can be performed. In this way, the effectiveness of braking is increased as the dimension of brakes remains the same, even if a higher specific investment and a higher cost of machining are necessary. Conversely, by turning with accuracy, a lower precision on flatness and parallelism of friction surfaces will be achieved, even if the match with the braking device is improved by improving the life of the moulding friction. This last solution is cheaper and the selection between the two options depends on the product's targets.
2. Based on the above-mentioned project analysis, suitable industrial processes are chosen, assuring that:

- machines and equipment will be robust and accurate to guarantee "intrinsic reliability";
- machine's gears and leverages will assure the necessary "repeatability" of the process, by applying the necessary self-compensation systems;
- tools and gigs will assure necessary "precision" for each operation of the productive cycle;
- measuring techniques and tools will be coherent with "Process Capability" requirements;
- an ad hoc monitoring system for checking process parameters will be adopted, with an effective visual management of diagnosis indicators easy to be read by machine operators;
- operators and maintenance operators will be properly trained to check and adjust parameters, substitute spare parts and worn tools and perform maintenance interventions necessary to maintain the "process capability" over time (according to the organizational logic discussed in Chap. 5).
We emphasize that eventual non-fulfilment in applying the above preventive criteria will lead to problems in obtaining and maintaining the values of "SIGMA" necessary to assure product quality, with "zero defects" in final product and the lowest number of rejects possible.
On the subject of *quality prevention,* let us lay out the modern techniques (CAD/CAE/CAPE) recommended in the design phase for optimizing the manufacturing system's "process capability":

(a) **Transformation and Simulation of Element Forming Processes**
Starting from tri-dimensional CAD drawings of mouldings and tools, by processing specific physical/chemical data of materials during the transformation phase, it is possible to simulate the filling of forms and the solidification and shrinking of materials, examining the influence of the process parameters. In this way, it is possible to adjust moulding drawings, relative thermal power supply and conditioning schemes to obtain the best productivity results and assure the desired quality levels. These virtual techniques are almost consolidated for processes of steel gravity casting

and polymer injection and injection-compression. Other satisfactory solutions are available for the forging of steel components and extrusion processes, as well as for forging processes by sintering.
Conversely, satisfactory solutions for thick sheet metal or aluminium stamping by pressing are still not available, this for the dispersion due to the elastic behaviour of metals during the process (mostly due to thickness variation and to gardening during lamination). But even for this technological area, significant improvements will be introduced in the next few years, thanks to new research.
By applying the above-mentioned virtual techniques, time and costs required for experimentation and adjusting of dies are reduced, defining in advance the "process capability" level required and obtainable.

(b) **Assembly Processes Simulations**
Starting from elements to be assembled and tools and jigs for framing and positioning tri-dimensional drawings (CAD), the kinematic functioning of the process may be represented virtually, so that it is possible to:

- avoid and prevent interference or excess of gaps in assembly and adjustment of parts (system called ROBCAD, DIGITAL MOCK-UP, etc.);
- assure the possibility of the proper assembly of components and subassemblies, complying with the ergonomic criteria for working and the use of partner tools.
The above virtual verifications are very useful for body part subassembly design activities and for manufacturing engineering plan elaboration, because they allow for the minimization of time and costs for experimentation necessary to optimize the product's functionality and styling.
During these project verifications, it is important to analyse in terms of probability the effects of the "chain of tolerance", trying to assign for each part and component the right precision and exact tolerance, selecting, when needed, the elements produced in matching classes.
For mechanical assembly in power-train systems, we mention the following examples of selected preparation of components, after having divided the components into classes of high precision of centesimal order: piston/rod matching, application of selected half-bearings to optimize the driving shaft pin/sedge diametrical gap. These selections lead to specific measuring, recognition and automatic assembly techniques.

(c) **Software Programs for Process Automation Simulations**
To develop *hard automation manufacturing systems*, it is necessary to verify the reliability of software programs during the design phase, through the simulation of "time sequence diagrams", by introducing process variables that can introduce instability in normal functioning.
In this way, machinery and tool design is optimized, shortening time needed to try out systems during productive start-up.

8.4 Quality and Reliability Assurance Techniques in Product Design Phase

To reach the level of quality demanded in the automotive industry, it is fundamental to apply the following analysis, necessary for "preventing" eventual defects in the final product or process failures that could determine production rejects or reworking operations:

(a) ***Product FMEA and Design Review***
During the design phase, the probable "failure modes" of a product's components and subsystems are detected by predicting them by induction method, on a statistical basis. As a consequence, the necessary "design reviews" are set to strengthen the critical part, assuring the required reliability. In the meantime, it is important to search for a lean solution by eliminating eventual redundancies in terms of weight, cost and maintainability.

(b) ***Process FMEA and Simulation Methodologies***
In both the definition of the manufacturing engineering plan phase and the design of machines and specific equipment phase, the probable process failure modes with critical consequences for product quality must be determined. As a consequence, the necessary countermeasures in terms of virtual and experimental verifications are taken, so that the proper level of "capability" in the transformation processes will be consistent with product objectives (range of tolerance, Cp and Cpk values), in an attempt to eliminate eventual redundancies that cost money and investments.

(c) ***Experiment Techniques for Products and Process Control***
In relation to a project's technical data and process parameters considered "important", based on research of the above points (a) and (b), the experimentation plans for products and process control are established through statistical methods.

Also dealing with methodologies for the achievement of excellence in quality results, we mention the "*Six Sigma*" system, particularly widespread in compartments with technologically complex products, with high impacts on safety and environmental aspects. It was conceived in the USA, and adopted by NASA to solve the biggest problems of reliability in the aero-space industries.

Further on, in the '70s and '80s, other companies, including Motorola and General Electric, successfully adopted the "SIX SIGMA" application for other important industrial applications: civil aeronautics, railway transportation, thermal-nuclear power stations, chemical processes, etc.

The logic of "SIX SIGMA" consists in containing the possible dispersion range of process variables that influence product and process reliabilities at a minimum level, with the aim of reducing the probability of failures to zero or to parts per million.

In recent years, "SIX SIGMA" has also been applied to some automotive industries: it effectively integrates the preventive methodologies mentioned in the previous points (a), (b) and (c) by establishing a systematic link between the several company departments engaged in management of industrial processes and after sales customer assistance, to obtain the best results for Quality and Safety.

8.5 Total Quality Management and Continuous Improvement Approach

First of all, we would like to note that, for quality management in industrial processes, it is fundamental to apply statistical methods and some basic quality tools, also known as *7 Basic Quality Tools.*

1. check sheet: this consists of checklists to be applied for verifying if specific conditions regarding quality have been applied during transformation processes;
2. bar graph (histogram) elaborations: for descriptions of critical defects, in terms of frequency and gravity;
3. Pareto analysis: starting from histograms, it is possible to prioritize the defects/phenomena by calculating the cumulate curve and by ordering them decreased (see example on the next page);
4. cause and effect diagrams: these consist of a fish-bone diagram by which the four main factors of analysis of causes (man, method, machine, material) are grouped separately and connected through a cause-effect relationship (see following pages for more details);
5. analysis of scattering of process variables, according to Gaüss's theory;
6. control chart: this is a graph or chart with limit lines, called control lines (Upper Control Limit, Lower Control Limit and Central Line); the purpose of drawing a control chart is to detect any changes in the process that would be evident by any abnormal points listed on the graph from the data collected;
7. N_p attribute control chart: when items are compared by some standard and then are classified as to whether or not they meet that standard; The Np control chart is used to determine if the rate of nonconforming product is stable, and will detect when a deviation from stability has occurred.

To gather actions for improvement in order of priority, it is important to detect problems by classifying them by gravity and relevance, searching for the root causes through theoretical analysis and experimental research. In Chap. 5, we examined how to proceed with preventive actions for equipment failures due to breakdowns. In the same way, it is possible to improve the quality level of products and processes.

The easier and more effective statistical method consists of elaborating a diagram like the one shown below: on the x axle, "critical phenomena" in order of gravity (bar diagrams) are grouped by type; on the y axle, the relative impacts are shown. It is then possible to draw the cumulative curve of produced effects (Pareto curve).

According to the diagram in Fig. 8.4, by selecting and approaching three phenomena out of fifteen, the improvement obtainable is equal to 66 % (i.e., reducing the impact of defects by 2/3); if we extend intervention to 50 % of the phenomena, the improvement obtainable increases to 90 %.

We emphasize that the Pareto diagram technique is useful not only for quality and function reliability phenomena, but also for detecting priorities for positive phenomena such as pushing sales for product with a better profit margin, using the same productive capacity.

After this introduction on the importance of statistical methods, let us deal with the most popular organizational methodologies applied by companies to achieve excellence in quality and productivity.

To generate "new ideas for improvement", an enterprise's internal and external resources are involved, from internal resources up to the provider of engineering services, equipment and components, through the application of techniques of "continuous improvement." These have been adopted in several industrial industries by the best competitors for the achievement of excellence, notably, in the automotive industry, Toyota, which, in the last 40 years, has realized a very efficient production and industrial system (the so-called Toyota Production System), based on the logic of Kaizen for continuous improvement of their standard.

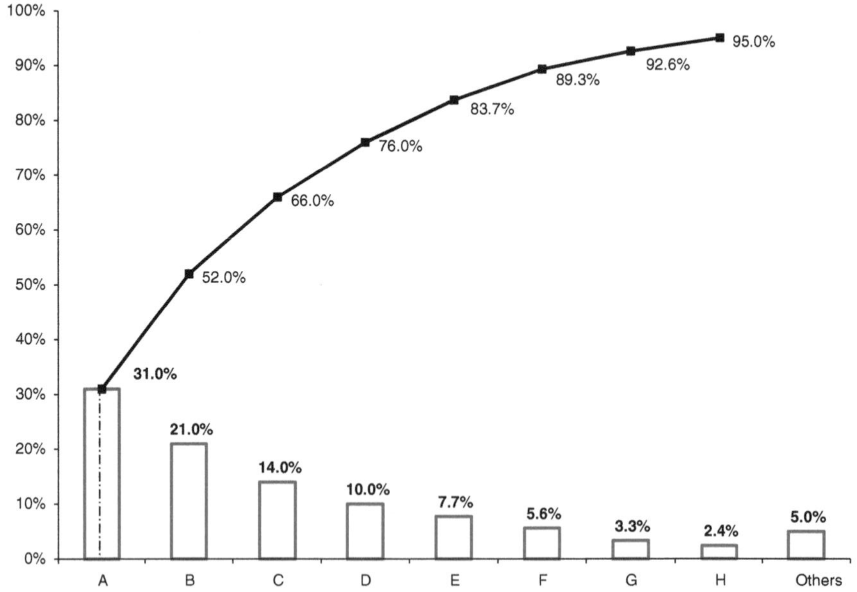

Fig. 8.4 Pareto diagram example

Some of the main logistics and organizational methodologies upon which the modern industrial organizations are based belong to the *Problem Solving* approach and involve the following basic tools:

- **5 W 1 H**
  This is a very simple tool for describing a situation or phenomenon by using six simple questions; the Five Ws and one H is a common concept used in journalism, police investigations and other endeavours involving research and basic information-gathering.
  It is a formula for getting the "full" story on something. The maxim of the Five Ws and one H is that, in order for a report to be considered complete, it must answer a checklist of six questions, each of which is comprised of an interrogative word:

  1. Who?
  2. What?
  3. Where?
  4. When?
  5. Why?
  6. How?

In the end, it is possible to consolidate all the elements and formulate a good description of the phenomena to be analysed.

- **Brainstorming**
  This consists of exchanging ideas openly, within a small cross-functional group of workers, obtaining a full analysis of all aspects, a good discussion on suggestions for improvement, and a complete sharing of solutions adopted; the first phase of brainstorming must be conducted without rejecting any proposal, while the next phases are oriented towards verifying and checking opportunities to run the effective countermeasures. It represents the first step of cause analysis in problem solving.
- **Cause and Effect Analysis and the Ishikawa diagram**
  The Ishikawa diagram (also known as a fishbone diagram or cause-and-effect diagram) shows the causes of a certain event. A common use of the Ishikawa diagram is in product design, to identify potential factors that may cause an overall effect, but can be easily applied to other phenomena, such as defects and other failures in manufacturing processes. Kaoru Ishikawa proposed the Ishikawa diagrams in the 1960s, pioneering quality management processes in the Kawasaki shipyards, and, as a result, becoming one of the founding fathers of modern management. The diagram was first used in the 1960s, and is considered one of the seven basic tools of quality (see above).
  Most Ishikawa diagrams have a box at the right-hand side, where the effect to be examined is written. The main body of the diagram is a horizontal line from which stems the general causes, represented as "bones." These are drawn towards the left-hand side of the paper and are each labelled with the causes to be investigated, often brainstormed beforehand and based on the major causes

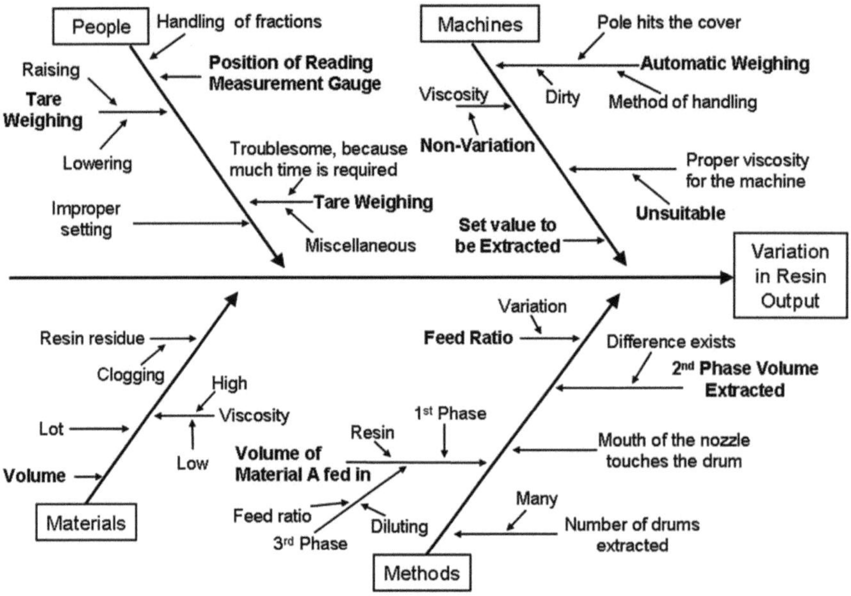

Fig. 8.5 Cause-effect fish bone diagram example

listed above.

Off each of the larger bones, there may be smaller bones highlighting more specific aspects of a certain cause, and sometimes there may be a third level of bones or more. These can be found using the '5 Whys' technique (see below). When the most probable causes have been identified, they are written in the box along with the original effect. The more populated bones generally outline more influential factors, with the opposite applying to bones with fewer "branches." Further analysis of the diagram can be achieved with a Pareto chart.

Figure 8.5 uses, as an example, the effect of "the variation of quantity in resin output" of equipment; after a brainstorming session, all potential causes with different levels of relations, are displayed in the Ishikawa diagram.

- **5 Whys**

  The 5 Whys is a question-asking method used to explore the cause/effect relationships underlying a particular problem. Ultimately, the goal of applying the 5 Whys method is to determine the root cause of a defect or problem.

  The technique was originally developed by Sakichi Toyoda[1] and was later used within the Toyota Motor Corporation during the evolution of their manufacturing methodologies. It is a critical component of problem-solving training

[1] Sakichi Toyoda (February 14, 1867–October 30, 1930) was a Japanese inventor and industrialist. He was born in Kosai, Shizuoka. The son of a poor carpenter, Toyoda is referred to as the "King of Japanese Inventors." He is often referred to as the father of the Japanese industrial revolution. He is also the founder of Toyota Industries Co., Ltd.

delivered as part of the induction into the Toyota Production System. The architect of the Toyota Production System, Taiichi Ohno,[2] described the 5 Whys method as "...the basis of Toyota's scientific approach ... by repeating 'why' five times, the nature of the problem as well as its solution becomes clear."

- **Plan Do Check Act**

  This is the basic approach for all problem-solving activities; it was invented by Edward Deming[3] and is also known as the Deming Cycle. It works in the following way:

Plan	Establish the objectives and processes necessary to deliver results in accordance with the expected output. By making the expected output the focus, it differs from other approaches in that the completeness and accuracy of the specification is also part of the improvement
Do	Implement the new processes
Check	Measure the new processes and compare the results against the expected results to ascertain any differences
Act	Analyse the differences to determine their cause. Each will be part of either one or more of the P-D-C-A steps. Determine where to apply changes that will effect improvement. When a pass through these four steps does not result in the need to improve, refine the scope to which PDCA is applied until there is a plan that involves improvement.

- **Continuous Improvement and Kaizen Activities**

  Kaizen (Japanese for "continuous improvement") is a Japanese philosophy that focuses on continuous improvement throughout all aspects of life.

  When applied to the workplace, kaizen activities continually improve all functions of a business, from manufacturing to management and from the CEO to the assembly line workers. By improving standardized activities and processes, kaizen aims to eliminate waste. The logic applied is very close to the PDCA method mentioned above, and considers the "actual standard" as the starting point for improvement to create a "new standard", which becomes the arrival point of the kaizen activity, as displayed in Fig. 8.6.

Using this approach, it is possible to improve through small steps, being sure not to lose the benefit achieved from the improvement activities over time, because

[2] Taiichi Ohno (February 29, 1912–May 28, 1990) is considered to be the father of the Toyota Production System, also known as Lean Manufacturing. He wrote several books about the system, the most popular of which is Toyota Production System: Beyond Large-Scale Production. Born in Dalian, China, and a graduate of the Nagoya Technical High School, he was an employee first of the Toyoda family's Toyoda Spinning, then moved to the motor company in 1943, and gradually rose through the ranks to become an executive.

[3] William Edwards Deming (October 14, 1900–December 20, 1993) was an American statistician, professor, author, lecturer and consultant. In Japan, from 1950 onwards, he taught top management how to improve design (and thus service), product quality, testing, and sales (the last through global markets) through various methods, including the application of statistical methods.

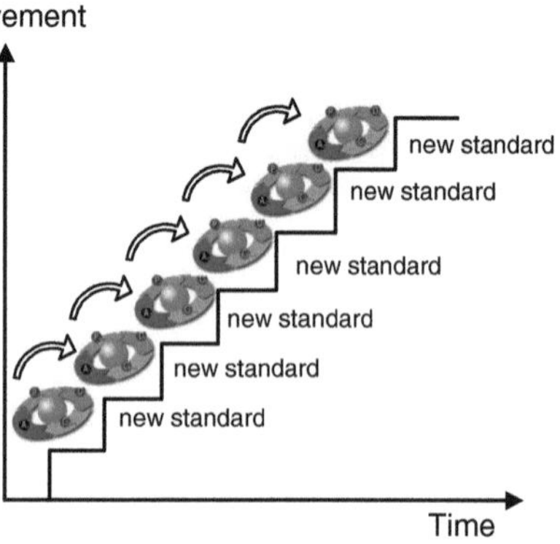

Fig. 8.6 PDCA cycle and new standard achievement through continuous improvement logic

of the base standard created (Fig. 8.7a). Conversely, by improving in big steps (radical technological innovations), there is a risk of regression if no standard has been created to support the innovation (Fig. 8.7b). The ideal condition is to combine the big improvements (less frequent) with a consistent application of smaller ones (continuous improvement, Fig. 8.7c).

Kaizen was first implemented in several Japanese businesses during the country's recovery after World War II, including Toyota, and has since spread to businesses throughout the world.

The Toyota Production System is known for kaizen, where all line personnel are expected to halt production on their respective moving lines in case of any abnormality and, along with their supervisor, suggest an improvement to resolve the abnormality in order to initiate a kaizen

The cycle of kaizen activity can be defined as:

- standardize an operation
- measure the standardized operation (find cycle time and amount of in-process inventory)
- gauge measurements against requirements
- innovate to meet requirements and increase productivity
- standardize the new, improved operations
- continue cycle ad infinitum

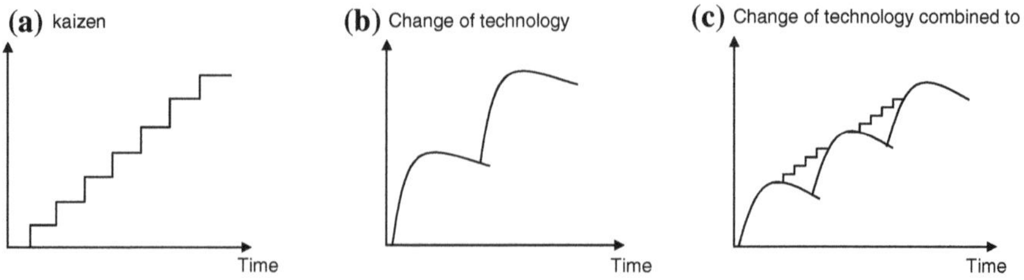

Fig. 8.7 Different approaches to improvement

Considering the entity and complexity of the phenomena, kaizen activities can be different in terms of time, resources and dedicated tools, distinguishing more levels of kaizen activity.

Other important techniques that can be used in approaching Continuous Improvement are:

- **Benchmarking**
  It is a good approach to introduce new ideas and stimulate new energies for improving organizations and methodologies that can really shore up industrial competitiveness.
  First, products and processes are identified that are considered "critical" and for which it is a priority to introduce improvements in relation to customer expectations and competitors' scenarios. Then, organizations (belonging to the same or to other enterprises) with the wherewithal to lead the research and establish a benchmark are identified and contacts are made to arrange for visits and exchange experiences, starting up analysis ad hoc to identify the best practices to be adopted.
- **Quality Circles**
  These are volunteer groups composed of workers who meet to talk about workplace improvement, and then make presentations to management communicating their ideas, especially relating to quality of output, in order to improve the performance of the organization and to motivate and enrich the work of the employees. Typical topics are improving occupational safety and health, improving product design, and improvement in the manufacturing process. The ideal size of a quality circle is eight to ten members.

Quality circles were first established in Japan in 1962, and Kaoru Ishikawa[4] has been credited with their creation. The movement in Japan was coordinated by the Japanese Union of Scientists and Engineers (JUSE). The use of quality circles then spread beyond Japan. Quality circles have been implemented elsewhere, such as in India where they have proven useful to the educational sector, and QCFI (Quality Circle Forum of India) has promoted such activities.

Considering all of the above-mentioned logistics and, in particular, the Continuous Improvement approach, it is important to introduce the concept of *Total Quality Management*, and the typical way to approach quality in industrial processes:

(1) **Reactive approach: maintenance of a constant quality level**
This consists of checking quality on the products at different points of the production processes (from raw material to final products) and reacting to problems detected with specific countermeasures so that a constant quality

---

[4] Kaoru Ishikawa (1915–1989) was a Japanese University professor and influential quality management innovator best known in North America for the Ishikawa or cause and effect diagram (also known as Fishbone Diagram) used in the analysis of multiple cause phenomena.

level may continue to be achieved. This is a traditional approach in which quality is obtained through inspection and checking results on the product.

(2) **Preventive approach: continuous improvement of actual standards**
By using the inspections and checks on the product and introducing specific inspections throughout the process, it is possible to use kaizen activities to improve the quality standards to fit better with customer requests. With this approach, it is very important to orientate to benchmarking activities for quick improvement of the best practices and, as a consequence, the quality standards of products and processes. In this logistic, the focus is on research of the condition of "zero defects" to be set within the industrial processes (design, industrialization and manufacturing), shifting the quality approach from inspection of the product to control of the processes.

(3) **Proactive approach: anticipating the generation of errors and defects**
At this level of application, the conditions for zero defects have been clearly set and the kaizen actions are directed to prevent and improve the risks of generating defects and errors that could also lead to an increase in non-quality costs.

In modern and advanced industrial applications, where there are more products and service lines, all three approaches could exist, even if it were more beneficial to move from a preventive approach to a proactive one; to adopt the right approach, the benefit to cost ratio and the customer's perception of quality should be considered.

The typical conditions of defect rate improvement consequent to the improvement of actions are shown in Fig. 8.8.

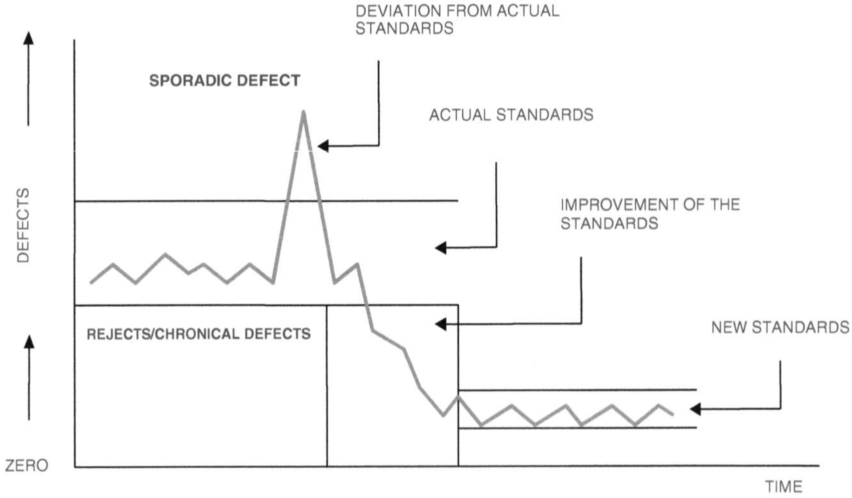

Fig. 8.8 Defect rate index and improvement actions effect

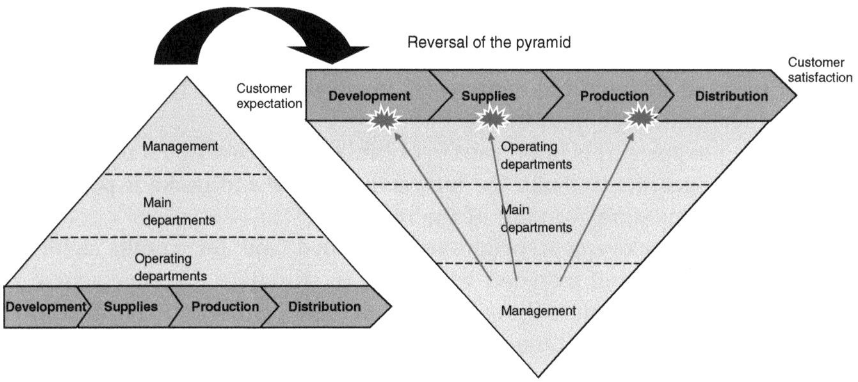

Fig. 8.9 Organization pyramid reverse through continuous improvement approach

In using a continuous improvement approach, enterprises must also adopt an effective organization in which the right commitment will follow a "top-down" line and proposals will follow a "bottom-up" line, reversing the classic pyramidal organization, according to what is displayed in Fig. 8.9. Continuous Improvement makes it possible to overcome the classical organizational template of the company, where all the departments address losses in the value chain.

On the Value Chain line, the small kaizen team contributes to reduce rejects and waste, while the other staff departments contribute with their support to the continuous improvement of the standards. The success can be directly measured in real benefits by using Key Performance Indicators and also translated into cost savings, according to Finance Department cost accounting rules, while the best performing teams are rewarded in terms of social involvement and sometimes also financially.

Through an organization like this one, Toyota has designed its own Production System, known as *TPS* (*Toyota Production System*), in which the main logics of Total Productive Maintenance (TPM), Total Quality Management (TQM), Total Industrial Engineering (TIE) and Just in Time (JIT) can be applied, with the involvement of all employees in the industrial organization, from top management to the workers. TPS has become a real benchmark over time, not only for automotive industries, but for all other industrial compartments, inspiring several lean manufacturing models of industrial organization based on these principles for achieving a world class level in manufacturing applications and beyond.

To conclude this section, it is useful to summarize the main concepts mentioned above:

- **Continuous Improvement activities must not be considered as "una tantum initiatives", but as "normal activities" constantly necessary to achieve and maintain the best quality levels for guaranteeing the brand image and the customers' satisfaction.**

- **At the shop floor level, lean organizational procedures should be established, so that it is possible to focus improvement activities according to priorities and through the employees' involvement, clearly demonstrating Top Management's commitment.**
- **For these purposes, it is important to organize cross-functional teams, made up of a few people, to make the interaction easier and make it possible to collect the suggestions of each of the members.**
- **Awards for improvement proposals generated and for results achieved must be weighed and assigned with care, to stimulate the engagement of each worker and promote the collaboration of all employees: certification to the team, public presentations of results, prizes, not necessarily financial, development of professionalism and opportunities according to merit.**

Chapter 9
Value Creation and Final Considerations

Let us now reconsider the entire book, underlining correlations between the different sections and providing a final synthesis.

After having dealt in Chap. 1 with industrial policies that determine the strategic assets of manufacturing activities, we explored in Chap. 2 how the industrial development of a product is managed, according to sales targets and customer expectations.

In Chap. 3, we discussed methodological criteria applied to setting production capacities in a competitive scenario, assuring the necessary degree of utilization of investments.

Chapters 4 and 5, strictly in relation to Chap. 3, dealt with the organizational criteria of labour and management of technical systems, with an eye towards quality targets, productivity and flexibility.

In Chap. 6, we examined the "logistic process" , from supply sources to production flow management, up to delivery of final products, emphasizing certain strategic factors such as the service level toward the commercial network and the order lead time.

Chapter 7 was dedicated to the Customer/Supplier relationship throughout the Procurement activities, with reference to the modern policies of global sourcing to assure effective cooperation between the enterprises that are engaged in product development and productive activities, on a competitive "worldwide" basis.

Finally, in Chap. 8, we considered the main modern methodologies applied to obtain "quality for industrial processes", according to continuous improvement logic, finalized to empower the company's brand and achieve/maintain customer fidelity.

We have attempted to offer the reader a "wide ranging" vision of automotive industrial processes.

To conclude, we think it is important to focus on the contribution asked of Production Management and Industrial Co-operations in "creating value" for an enterprise.

Let us remember that companies' business units create value when the margins obtained from sales of the products/services are positive, sufficiently covering the net invested capital and the burden, rewarding entrepreneurial risk.

M. Gobetto, *Operations Management in Automotive Industries*,
Springer Series in Advanced Manufacturing, DOI: 10.1007/978-94-007-7593-0_9,

It is clear that, at the same value and price of sold products and services, if we reduce production and commercial costs (including warranty expenses), the gross profit margin increases proportionally, as a determinant factor to "create value". For this purpose, during the design phase, it is important to balance the product's value with its cost ratio, to obtain, during the operative phase, the highest efficiency along the entire *supply chain*.

Every initiative of renewal of products and empowering/improvement of productive capacities must be well-planned, in relation to the profit targets and the obtainable operative margin, assuring that the financial effort will be remunerative. The "business planning" criteria and the methodologies for investment profitability analysis are not included in this text, being matters of Enterprise Economy. By the way, we think it important to emphasize that an effective setting of production processes, with the right selection of technologies applied, can contribute to generate value for enterprises.

The scheme in Fig. 9.1 synthesizes this topic, with reference to the items already discussed in Chap. 1 and to the strategic characteristics for the competitiveness of enterprises.

Another recurrent "leit motif" in all chapters is the concept of **economy of scale**. As we saw in Sects. 1.2 and 2.1, this is strictly related to the strategic plan for product range and the productive capacities. The standardization process makes possible the horizontal expansion of components and working means,

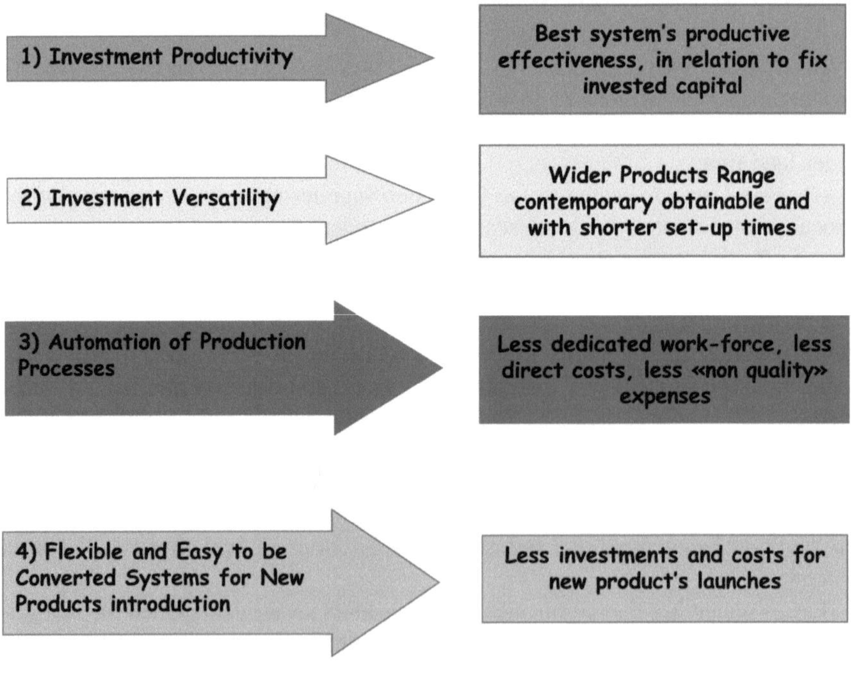

Fig. 9.1 "Production system" prerogatives that contribute to create value for the enterprise

Table 9.1 Economy of scale actions and effects

Action	Effects
Standardize industrial products and processes on a competitive basis	—Lower R&D expenses
Develop further innovation through strategic alliances with other companies	—Lower investments
Set optimal manufacturing systems in terms of degree of productivity and utilization	—Lower variable transformation costs —Lower incidence of depreciations and fixed costs
Concentrate supplies on highly competitive "worldwide" sources	—Lower variable costs for purchasing of components and consumable materials
Integrate information technology systems	—Lower incidence of fixed company costs
Speed-up logistic flows along the plants and the entire *supply chain*	—Lower working capital invested

so that a more intensive utilization of resources and investments can be achieved. When sales and profit do not reward investments, possible "co-makership" alliances should be researched, to offer proper industrial synergies.

Let us review the actions that promote "economy of scale" in the scheme in Table 9.1.

These positive effects can be obtained only when products and services for the relative markets meet expected commercial success.

The "economy of scale" is promoted through the "learning curve" that is progressively cumulated over time (for this argument, see Sect. 8.1)**. For this purpose, it is very important to take care of the "company's technical know-how" and promote, for specific needs, professional training for updates.**

For effective management of industrial activities, it is necessary to have a good control system for finance and economics. For this purpose, managers should assure that data available will be qualitatively good (complete and true) and quickly transmitted to the modern information technology systems used by the company. It is for this that, in every chapter of this text, we have shown the characteristic parameters of cost of the analysed processes, specifying the relative key performance indicators that can be used for operative control.

Industrial activities managers (R&D, Manufacturing, Purchasing, Supply Chain…) must evaluate and analyse data processed by the company's accounting system in advance to detect causes of variation, according to their positions in the enterprise's organization.

In very large companies, all the different departments must interact, under the coordination of Central Staff for Economic and Financial Control, to prepare budget and strategic plans according to the specific operative targets in the medium-short term and to Top Management's strategic plans.

It is on these organizational logistics, adopted by the best-performing international enterprises, that the modern criteria of managerial control and delegation stand.

Appendix
Production Management Basics

A.1 The Product

A.1.1 Product Development

Every enterprise exists in an environment in which economic conditions and society are continuously changing and evolving. Therefore it is important to attempt to forecast what the future of the environment will be in its different aspects, and to consider the environmental constraints that might limit the behavior of the enterprise (Fig. A.1).

The principal aspects to be considered in this type of analysis are:

- **The demand.** Every enterprise must consider the current and the expected characteristics of its market. Such studies must be aimed at appraising the future possible market background, assessing the possible effects of decisions already taken, anticipating decisions that might consider only single environmental hypotheses, and finally contributing to the development of the enterprise into an expanding organization.
- **The offer.** It is fundamental to examine the behavior of competitors, to consider their composition, their market share, their sales policies, and their financial position. Such factors must be carefully sifted to determine the degree of their vulnerability and therefore to take advantage of this knowledge and analysis.
- **The technological process.** Considering the numerous innovations in every field of business activities, no enterprise can avoid technological progress in increasing their own competitiveness.
- **Rules and regulations.** An external constraint of primary importance for any enterprise is the effect of public regulation on private entrepreneurial activities. Many restrictions, such as hygienic-sanitary regulations, environmental protection regulations, inland trade and monopoly regulations, can limit the freedom of the enterprises themselves.

M. Gobetto, *Operations Management in Automotive Industries*,
Springer Series in Advanced Manufacturing DOI: 10.1007/978-94-007-7593-0,

Fig. A.1 Enterprise and external implications

The development of a new product involves preliminary research, aimed at formulation of the "concept[1]" and definition of the "mission" in the market. The qualitative and quantitative analysis of demand defines the sales objectives, considering the market positions of competitors. In such a way an expected production level is planned, on a sufficiently wide temporal horizon. This leads to activity phases related to executive management and to realization of the production process. The cycle of the product development concludes with the beginning of production and introduction into the market.

In the organizational structure of the enterprise, the **marketing** division has the strategic role of maintaining an eye on the market, constantly observing the behavior of competitors and seeking new opportunities. All the industrial divisions (R&D, management of production, etc.) have to collaborate with marketing to assure the best results, and to produce "value for the enterprise". In Table A.1, the necessary operations for optimizing the planning of the industrial activities are generally summarized.

A.1.2 The Product Life Cycle

A product's life cycle (PLC) can be divided into several stages characterized by the revenue generated by the product. If a curve is drawn showing product revenue over time, it may take one of many different shapes, an example is shown in Fig. A.2.

The life cycle concept may apply to a brand or to a category of product. Its duration may be as short as a few months for a fad item or a century or more for product categories such as the gasoline-powered automobile.

---

[1] The concept is the basic idea of the product. It determines the clients target, the needs the product should satisfy and the product's general characteristics (in terms of performances, functioning etc.).

Table A.1 Planning of industrial activities

1. Analysis of opportunities	Who buys? What does he buy? Why does he buy?
2. Selection of products and the services required by the market	Which product? For which market? Which differentiation?
3. Segmentation and selection of targets	To individualize and to select the market segments
4. Marketing mix: – Product – Price – Promotion – Distribution	To organize the necessary activities to "feed" the selected market segment
5. Control	To constantly verify if management follows the established plans
6. Evaluation of the results	To appraise if the objectives have been reached

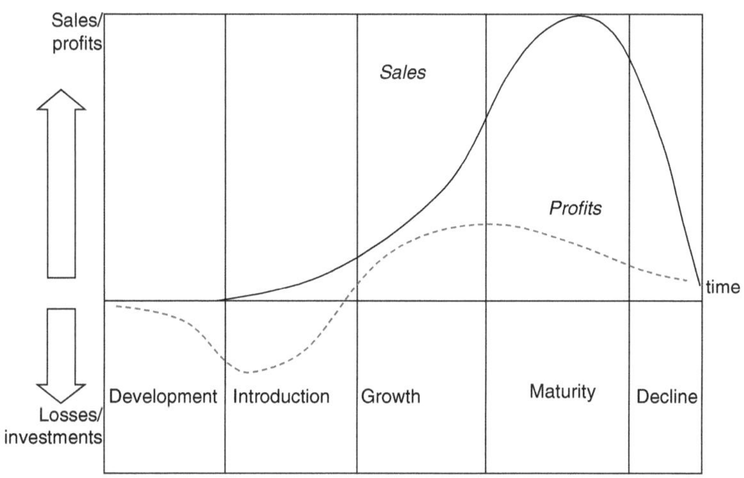

Fig. A.2 Product life cycle sales profit versus investments

Product Development

Product development is the incubation stage of the product life cycle. There are no sales and the firm prepares to introduce the product. As the product progresses through its life cycle, changes in the marketing mix usually are required in order to adjust to evolving challenges and opportunities.

Introduction Stage

When the product is introduced, sales will be low until customers become aware of the product and its benefits. Some firms may announce their product before it is introduced, but such announcements also alert competitors and remove the element of surprise. Advertising costs typically are high during this stage in order to rapidly increase customer awareness of the product and to target early adopters. During the introductory stage the firm is likely to incur additional costs associated with the initial distribution of the product. These higher costs coupled with a low sales volume usually make the introduction stage a period of negative profits.

During the introduction stage, the primary goal is to establish a market and build primary demand for the product class. The following are some of the marketing mix implications of the introduction stage:

- *Product*—one or few products, relatively undifferentiated.
- *Price*—Generally high, assuming a skim pricing strategy for a high profit margin as the early adopters buy the product and the firm seeks to recoup development costs quickly. In some cases a penetration pricing strategy is used and introductory prices are set low to gain market share rapidly.
- *Distribution*—Distribution is selective and scattered as the firm commences implementation of the distribution plan.
- *Promotion*—Promotion is aimed at building brand awareness. Samples or trial incentives may be directed toward early adopters. The introductory promotion also is intended to convince potential resellers to carry the product.

Growth Stage

The growth stage is a period of rapid revenue growth. Sales increase as more customers become aware of the product and its benefits and additional market segments are targeted. Once the product has been proven a success and customers begin asking for it, sales will increase further as more retailers become interested in carrying it. The marketing team may expand the distribution at this point. When competitors enter the market, often during the latter part of the growth stage, there may be price competition and/or increased promotional costs in order to convince consumers that the firm's product is better than that of the competition.

During the growth stage, the goal is to gain consumer preference and increase sales. The marketing mix may be modified as follows:

- *Product*—New product features and packaging options; improvement of product quality.
- *Price*—Maintained at a high level if demand is high, or reduced to capture additional customers.
- *Distribution*—Distribution becomes more intensive. Trade discounts are minimal if resellers show a strong interest in the product.
- *Promotion*—Increased advertising to build brand preference.

Maturity Stage

The maturity stage is the most profitable. While sales continue to increase into this stage, they do so at a slower pace. Because brand awareness is strong, advertising expenditures will be reduced. Competition may result in decreased market share and/or prices. The competing products may be very similar at this point, increasing the difficulty of differentiating the product. The firm places effort into encouraging competitors' customers to switch, increasing usage per customer, and converting non-users into customers. Sales promotions may be offered to encourage retailers to give the product more shelf space over competing products.

During the maturity stage, the primary goal is to maintain market share and extend the product life cycle. Marketing mix decisions may include:

- *Product*—Modifications are made and features are added in order to differentiate the product from competing products that may have been introduced.
- *Price*—Possible price reductions in response to competition while avoiding a price war.
- *Distribution*—New distribution channels and incentives to resellers in order to avoid losing shelf space.
- *Promotion*—Emphasis on differentiation and building of brand loyalty. Incentives to get competitors' customers to switch.

Decline Stage

Eventually sales begin to decline as the market becomes saturated, the product becomes technologically obsolete, or customer tastes change. If the product has developed brand loyalty, the profitability may be maintained longer. Unit costs may increase with declining production volumes and eventually no more profit can be made.

During the decline phase, the firm generally has three options:

- Maintain the product in hopes that competitors will exit. Reduce costs and find new uses for the product.
- Harvest it, reducing marketing support and coasting along until no more profit can be made.
- Discontinue the product when no more profit can be made or there is a successor product.

The marketing mix may be modified as follows:

- *Product*—The number of products in the product line may be reduced. Rejuvenate surviving products to make them look new again.
- *Price*—Prices may be lowered to liquidate inventory of discontinued products. Prices may be maintained for continued products serving a niche market.

- *Distribution*—Distribution becomes more selective. Channels that no longer are profitable are phased out.
- *Promotion*—Expenditures are lower and aimed at reinforcing the brand image for continued products.

Limitations of the Product Life Cycle Concept

The term "life cycle" implies a well-defined life cycle as observed in living organisms, but products do not have such a predictable life and the specific life cycle curves followed by different products vary substantially. Consequently, the life cycle concept is not well-suited for the forecasting of product sales. Furthermore, critics have argued that the product life cycle may become self-fulfilling. For example, if sales peak and then decline, managers may conclude that the product is in the decline phase and therefore cut the advertising budget, thus precipitating a further decline.

Nonetheless, the product life cycle concept helps marketing managers to plan alternate marketing strategies to address the challenges that their products are likely to face. It also is useful for monitoring sales results over time and comparing them to those of products having a similar life cycle.

A.1.3 The Bill of Material

The structure of a product, in terms of components and sub-components can be very complex. Such structure is represented in a list called bill of material—BOM. The BOM lists the type and the quantity of the sub-components that compose the product. The BOM is the list of materials and the necessary components to produce a determined end-product. The universally graphical way used to represent a BOM is a tree, as shown in Fig. A.3.

Given a certain independent demand for an end-product (P1 in the figure), the sub-components (P2, P3, P4, P5) demand can be derived from the BOM. In other words, the demand for sub-components depends on the "father" component demand. For instance a requirement of 100 pieces P1 produces a dependent demand of 100 pieces P2 and 200 pieces P3. The Bill of Material, besides being a

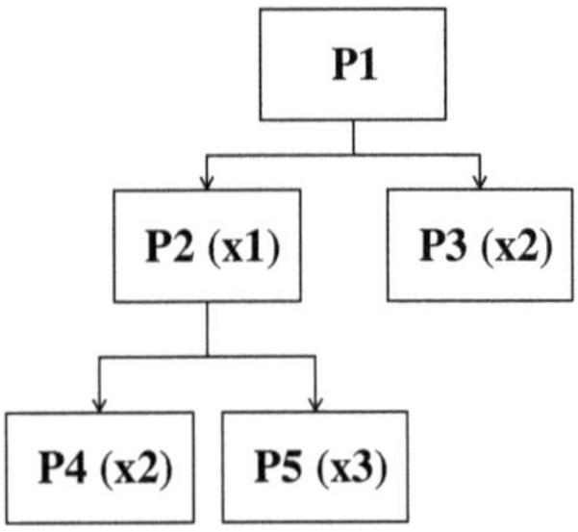

Fig. A.3 Bill of material tree structure

Table A.2 Bill of material matrix

		End-products							
		P1	P2	P3	P4	P5	P6	P7	P8
Common component	C1	4	–	2	1	3	2	–	2
	C2	3	3	2	1	1	1	2	1
	C3	2	2	2	3	1	4	5	2
	C4	3	1	3	1	–	1	2	4
	C5	2	–	2	3	4	2	3	1

list of product components, is a list structured in order to describe the necessary operational sequences for production of a determined article. The different levels of the BOM represent the different phases of product realization.

There are different types of BOMs: depending on the use, it can be a *planning* or *production* bill of material. The planning BOM, called also technical BOM, is usually completed with sketches that describe the components in geometric terms. Compared with the production BOM, it does not consider the phases of the machining and the material management, but it is very detailed in the components listing and description.

The **matrix BOM** in Table A.2 is a representation that connects the products to the components through the use of a matrix. It is used when different end-products are produced with the same sub-components.

The numbers inside the matrix point out how many components there are in a certain product.

A.2 Production Systems

A.2.1 Classification of How to Organize Production Systems

Figure A.4 summarizes the most common ways production systems are organized.

The way companies satisfy demand depends on the product degree of standardization/customization, the production process *lead time*,[2] and the kind of customers.

For complex, expensive and customized products (according to the customer's specification), production is started only after receipt of an order and the customer waits for the product delivery a necessary *lead time* for the product manufacturing. The name of this production strategy is ***make-to-order***, and products are often processed in small batches, because of the high variety of demand for expensive

[2] The **lead time** is the time between the production/selling order and the products delivery.

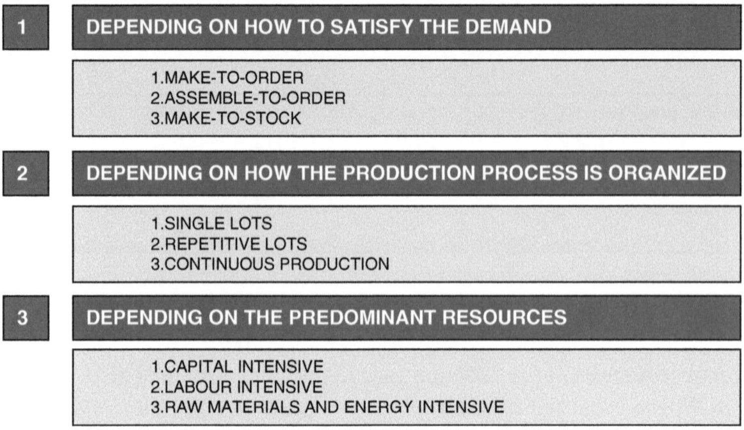

Fig. A.4 Organizational classification criteria for production

items, which makes maintaining inventories too costly. In other cases, the entire design of a product or a specific production process can be completely determined by the customer's order (*engineer-to-order*).

The opposite situation is ***make-to-stock***. In this case, standardized products are produced in advance (before they are demanded) and stored in the finished goods inventory (stock). The production amount and diversification are decided depending on the forecasted demand. Customers are immediately satisfied from the stock.

Make-to-stock and *make-to-order* are two extremes in the sense that with the first all production begins well before demand is received, whereas with the second approach production begins only after demand is known. Sometimes, neither a make-to-stock nor a make-to-order strategy is appropriate: consider a family of products which come in a large number of variants, obtained by assembling different combinations of components (which is the sum of the lead times related to the different levels of the BOM). Maintaining an inventory for each end item configuration is very expensive. On the other hand, the lead time necessary to produce the necessary components and to assemble them may be too long to adopt a make-to-order strategy. In this case, an ***assemble-to-order*** strategy may be adopted. We maintain an inventory of the components, and we assemble them on order. Assemble-to-order can be considered a hybrid strategy between make-to-stock and make-to-order. An extreme case of the make-to-order strategy is ***engineer-to-order***. Here accepting an order involves both the design and production of a highly customized and unique product.

The way to organize the production process depends on the processes technological characteristics and on the required production rate. Let us assume four different products (A, B, C and D) to be produced monthly in known quantities. Consequently a lot of units A, followed by a lot of units B, C and D are produced during the month. If a production is affected by large *set-up times*,[3] it can be convenient to perform a ***single-lots*** production, so that requirements for the whole planned period (typically a month or a week) are covered. On the other hand, if the set-up times are not so relevant, the single lot can be split into smaller lots, so that the production is organized according to a production mix which is consistent with the daily (or hourly) requirements of the single products. The result is that the production of small ***repetitive-lots*** of the different products is alternated during the month. Repetitive production is characterized by a production mix, i.e. by a set of production quantities that are periodically repeated; for instance, we might produce 100 items of type A, 150 of type B and 70 of type C per day. This mix is kept stable for a fair amount of time, and it is occasionally changed. In this way the production is "smoothed" in terms of qualitative composition of the products (*mixed model* production).This strategy minimizes the inventory level and the WIP and also reduces *lead-times*. Finally, ***continuous-production*** is suitable for large amounts of relatively simple products.

The kind of **predominant resources used in a production process** depends on the kind of activity, the technology, the target market. ***Capital-intensive*** companies use complex and expensive machines and systems. The influence of the amortization and the maintenance costs is predominant In general, *capital-intensive* companies are characterized by a high level of automation. For ***Labour-intensive*** companies, labour is the primary resource; typically, agricultural companies, clothing industries or service activities are *labour-intensive*. In the ***raw material and energy intensive*** companies, the major costs are due to the purchase of raw materials and the energy consumption; typical activities are metallurgic and petrochemical industries or power stations.

The automotive industry, includes both *labour-intensive* and *capital-intensive* activities.

A.2.2 Organizational Logics: Push and Pull

According to the organizational logic called **push** the direct materials (RM = Raw material) supply and the product manufacture are managed on the base of forecasts which *anticipate the demand.*

On the other hand, the organizational logic called **pull** provides a direct materials supply and determines the product manufacture to be managed, considering both the specific orders received by the commercial network and the substitution of products sold or used along the productive chain. It is a kind of production *tightly driven by the demand.* In *pull* systems materials or semi-finished products are "pulled" by the following production stages, instead of being

[3] **Set-up time** is the time needed for the changing of a machining process. During this time the machine is not working and machine equipment is substituted, generally causing a deceleration of the process when production is restarted.

"pushed" ahead, as it happens for the *push* systems. The production in every stage of the process is activated according to the needs of the following stages.

For *pull* systems inventories have a smaller relevance, because the production is driven by real demand, and anticipating the production is not necessary. The Finished Products (FP) are delivered directly to Customers, as shown in Fig. A.5.

A classic example of pull logic is given by the Japanese system kanban (literally "tag"): the sample of semi-finished or finished components causes the dispatch of a kanban to the previous production stages (inner departments or supplying establishments). The production goes on in relationship to the return of the kanban and the supply flow is thought to be "just in time", re-establishing every time the small existing stocks.

The two alternative logics (*push* and *pull*) have advantages and drawbacks. In a push system it is probable that the activities of the productive stages are not coordinated; if the lead times are overestimated, the previous production stages push the following stages earlier than needed, with consequent accumulation of intermediary stocks. Besides, if following production stages jam because of a critical event, the previous stages keep on producing according to the program. This doesn't happen in a *pull* system which produces only to replace the intermediary products already used. On the other hand, a *pull* system is not the proper system in case of productions with long lead times, when the clients demand should be promptly satisfied.

The choice among the two alternatives is made according to the production process features and depending on the policies of marketing. When it is possible, it is certainly preferable to adopt a *pull* logic, eventually turning to hybrid systems when the lead times are articulated in several manufacturing stages and supply sources. When the supply time required by clients (determined by market conditions) is less than the global time required for the complete productive process, some activities have to be done before the orders are placed.

A classical push and pull hybrid system is the so-called assemble-*to-order* (see Fig. A.6): the supply of base materials and standard products modules is planned based on forecast, whilst the final assemblage is done only on specific client orders. This kind of approach is a sort of border line between the two different production logics. Modern "integrated computing systems" are used to assure the

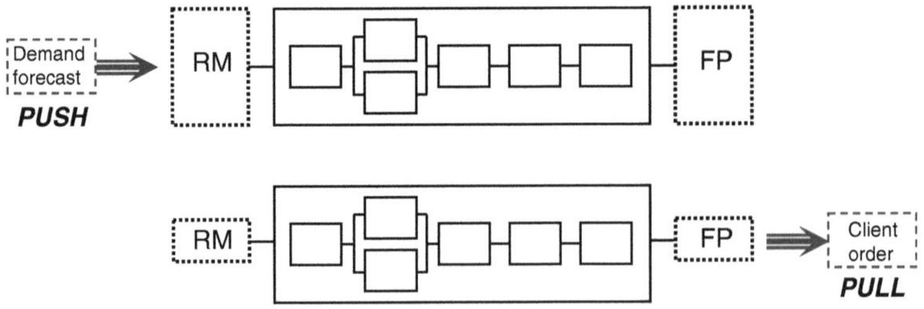

Fig. A.5 Push and pull logics

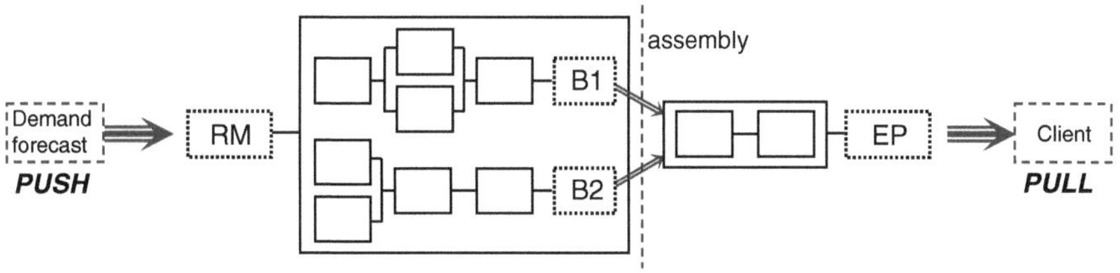

Fig. A.6 Assemble to order logic

global effectiveness of the planning process. In such a way the risks connected to the demand change are reduced, getting a more rapid reaction in the productive chain, reducing the circulating capital in the productive and distributive process.

A.2.3 Layout and Materials Flow

The various production scenarios call both for different management policies and different manufacturing system layouts. The physical layout reflects the structure of the manufacturing process (see examples in Fig. A.7).

a. **Product-oriented-layout.** In a low variety, high demand environment, it is worthwhile to organize the manufacturing system layout in order to maximize its efficiency[4] *(note 3)*: we obtain a **product-oriented-layout**, whereby products are allocated to dedicated production lines. Typically, different machines are connected in series and should be characterized by a high reliability for the reasons discussed in Chap. 3. The system throughput is corresponding to the Hourly Virtual Production (HVP), according to the definition given before.

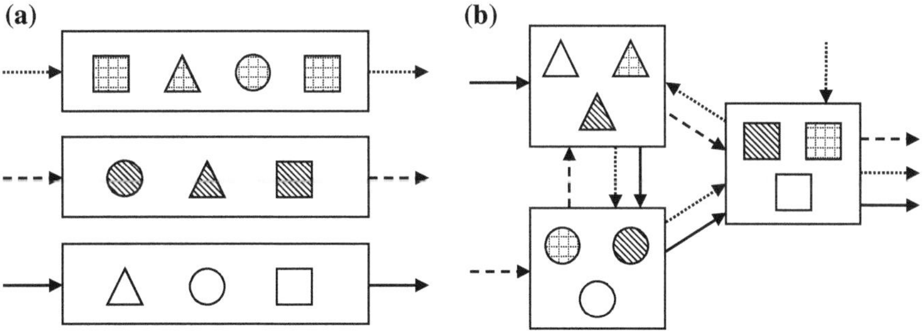

Fig. A.7 Layout logics: **a** Product-oriented-layout, **b** Process-oriented-layout

[4] **Efficiency** is intended as the ratio between the real production time and the total available time.

b. **Process-oriented-layout.** In a job shop environment, characterized by higher variety and lower demand, a larger degree of flexibility is required. It is natural to group machines according to their processing capabilities and functions to obtain a process-oriented-layout. A **process-oriented-layout** is more flexible than a product-oriented-layout, but material flow is scattered and difficult to manage; apart from material handling problems, there may be difficulties related to large set-up times.

Actually, there may be hybrid layouts including aspects from both the two previous layouts.

A.2.4 Performance Indicators

The performances of a production system are evaluated through some indicators:

- **Resources costs.** In production planning problems, it can be possible to vary the level of resources (machines and manpower), for instance through the use of overtime or contracting out a part of the production (subcontract).
- **Set-up related costs.** The set-up costs have a fundamental role both during the planning phase and the scheduling phase. Besides the costs, also the set-up times heavily influence other performance measurements. The set-up in fact reduces the productive ability, and this is particularly evident in case of bottlenecks. The costs/times of set-up can be **independent** or **dependent from the sequence**. In the first case, the time/cost of set-up depends only on the type of product to be produced; in the second case it also depends on the last type of worked product. For example, in the chemical industry, there are problems of chromatic or chemistry compatibility among different products: to produce the black varnish after the white is not problematic, while the other way round may need sensitively greater set-up costs/time.
- **Inventory level costs.** It is important to notice how inventory costs are negatively correlated with resources costs (machine and manpower) and with set-up costs. In fact, having an over-dimensioned production capability, it is possible to suit the production volume to the demand volume; with a reduced ability instead, it is necessary to accumulate stocks in the periods of low demand to be able to face following peaks of demand. Besides the presence of important set-up costs/times leads to the production of great dimensions lots, causing an increase of the inventory level. In a make-to-order context, not having end-products stocks, it could seem that such costs are not remarkable. In such case, however, there are some semi-finished products stocks (or work in process, WIP).
- ***Lead time.*** In a problem of stock control, the lead time is the time between the placing of a purchase or production order and the product's arrival. In case of scheduling it can be the time between the release of the materials into the warehouse and the completion of the production. In both cases the lead time is

tightly tied to the WIP, and it is not always exactly a-priori respectable. In the scheduling problems the term *flow time* is typically used.

- **Use of resources.** This is defined as the ratio between the time that a resource spends in production and the available time. The analysis of the resource's function is useful to spot bottlenecks and to improve system performances. It makes no sense to try to maximize the use of all resources; if one of them has a bottleneck, an excessive use of the preceding resource increases the WIP without any advantage. In general resources use and WIP are contrasting performance measurement.
- **Productive lilt or throughput.** This is defined as the number of parts produced in a given unit of time. The throughput is limited by bottlenecks and it is tied to the WIP level. It is possible to show that, under a general hypothesis the following relationship (known as law of Little) is worth:

$$WIP = throughput \cdot lead \quad time$$

- **Level of service.** A 95 % service level means that in 95 % of cases the order can be satisfied. The orders which cannot be satisfied because of a lack of materials constitute the *backlog* (or *backorders*) and they can only be satisfied late.
- **Handling Costs.** Material handling is particularly problematic in case of a process oriented layout. If the different departments are physically far from each other, the wait for material handling creates a WIP and the lead time increase.

The listed indicators are described with greater detail inside Chaps. 1 and 6, with specific reference to the productive and logistic trials in the automotive sector.

A.3 Criteria for Stock and Lots Management

A.3.1 Stocks Typologies and Relative Costs

There are many different typologies of direct materials stocks, depending on the progress of the integrated productive process and on the need for supply continuity:

- **End products,** these are products ready to be sold or delivered to clients.
- **Intermediary products and semi-finished products (WIP = work in process).** These are materials still in different work stations, as well as buffer stocks. Sometimes the production stages are geographically separated, even in different establishments. In such cases it is necessary to guarantee the necessary stocks to allow economic transport frequencies.
- **Raw materials,** incoming in the transformation process. Their supply flow is conditioned by the size of the economic transport lots.

- **Safety stocks.** They are useful to face critical situations, such as sudden changes of manufacturing program or production stops due to unexpected causes.
- **Seasonal stocks.** In case of products characterized by seasonal demand, it is necessary to prearrange opportune stocks in the period of low demand to be able to satisfy the demand in the period in which it is high.

There are many cost factors to be considered in stocks and lots management:

- **Inventory related costs.** These costs are due to capital immobilization, building rental, insurance etc.
- **Order related costs.** These costs are essentially related to products handling, machines set-up, bureaucratic files needed for the placing of orders.
- **Customer satisfaction related costs.** In case of end-products, these costs are related to client loss or to a brand image loss.

The last cost category, which is related to malfunctioning effects, is more difficult to quantify and it will be specifically described in Chap. 6.

A.3.2 Classification of Stock Control Systems

The criteria of stock management can be classified according to buying orders management.

- ***Periodic review:*** the decision of whether or not to buy is taken based on a periodical examination of inventory level. The period between two consecutive examinations is called the ***covering period***. Let "I" be the warehouse level found and "Q" the unknown quantity to be ordered. The buying order depends on two parameters: "S" the inventory target level (in general equal to the maximum quantity that the warehouse can stock), and "R" the restock level. This criterion can be schematized in the following way:

$$Q = 0 \text{ if } I > R,$$

$$Q = S - I \text{ if } I = < R.$$

The restock level (R) is useful to prevent too many small size orders, which may strongly affect the fixed costs. In the diagram (a) of Fig. A.8 the warehouse behavior is represented, when the periodic review criterion is applied, under the (unrealistic) hypothesis of *zero lead time*.

- ***Fixed order quantity (or continuous review):*** with this approach the order quantity Q is constant while the ordering moment is consequently determined. In this case the warehouse level must be constantly monitored, since a new stock is ordered when the inventory level goes under the point of restock R. The inventory level behavior is presented in diagram (b) of Fig. A.8.

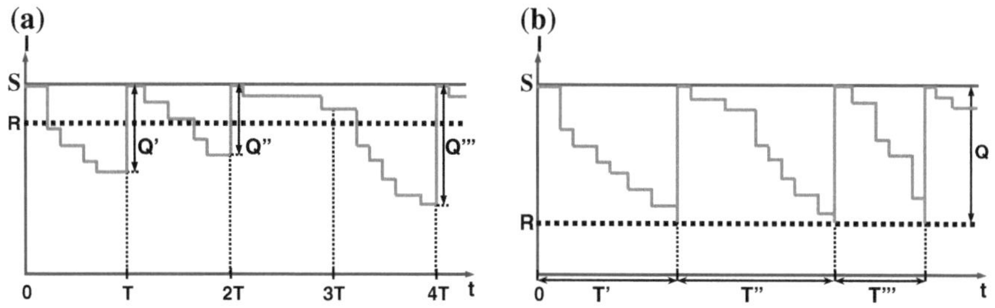

Fig. A.8 Re-ordering logics for stock management: **a** Periodic review, **b** Fixed order quantity

The lot sizing policies—used for lots management and for stock sizing—can be classified as static or dynamic. The **dynamic** models are able to treat time varying demands, while for **static** models the demand is assumed to be constant. The computational criteria can be deterministic or stochastic. In the first case the demand is considered defined, while in the second case it is considered a random variable, defined through a probability distribution. In the following we will deal only with deterministic models.

A.3.3 ABC Analysis and Hybrid Systems

In the previous section two classical alternative criteria of stocks management have been discussed, classifying them according to the kind of inventory inspection: periodic inspection (periodic review) and continuous inspection (continuous review or fixed order quantity). Both of them have some advantages and drawbacks: a system under periodic inspection is simpler to be managed, since it doesn't require a continuous monitoring of the stocks levels and makes it easier to control the different products order; on the other hand, a system under continuous examination eases the producer, since the ordered quantities are constant and not theoretically needed for safety stocks. Also in this case, it is possible to adopt hybrid criterions, so that a part of the products is managed in one way and the other part in the other way.

In many practical cases, the most of the overall value of the inventory stocks corresponds to a small portion of the stored products. The value of a stored product, in a determined period, can be expressed as the product between the relative demand rate and the unitary cost. A typical situation is represented in the graph of Fig. A.9.

It is commonly known that 20 % of a quantity of products represent 80 % of their total inventory value; such products are the A class product and are considered the most important. On the other hand, 65 % of the inventory products is made of C class products, whose value is only 5 % of the total inventory value and, therefore are not considered so important. There is then an intermediary class B, whose value

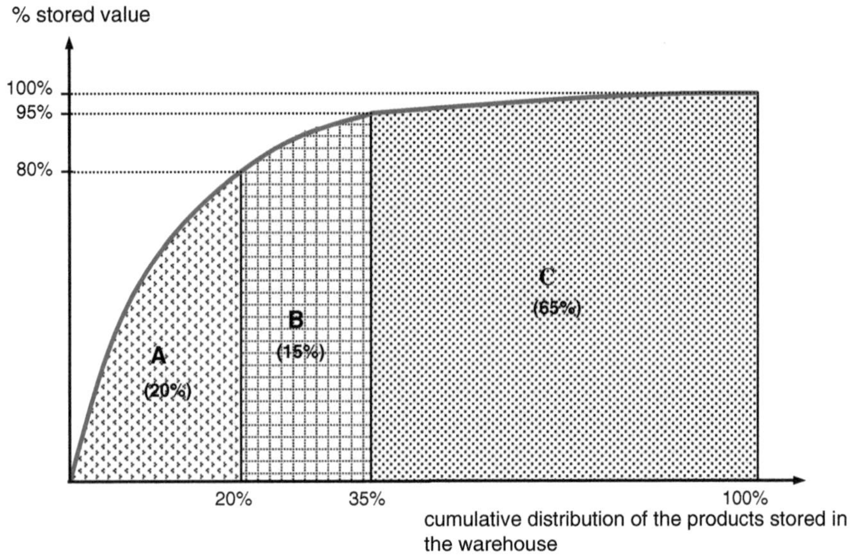

Fig. A.9 ABC material classification criterion

is 15 % of the total value. This criterion of stock value classification is known as ABC analysis.

The ABC approach allows minimizing the management costs of the stocks of different products, which are not modeled by the EOQ formulas discussed in the next section. The inventory verification and check has a certain cost; for this reason it is opportune to manage with the maximum care the A class products, assuring them a continuous and constant control. To the B class products is applied a control of middle level, while to the products of class C is applied an extremely simplified control.

A.3.4 Economic Order Quantity

The Economic Order Quantity (EOQ) is the number of units that a company should add to inventory with each order to minimize the total costs of inventory—such as holding costs, order costs, and shortage costs. The EOQ is used as part of a continuous review inventory system, in which the level of inventory is monitored at all times, and a fixed quantity is ordered each time the inventory level reaches a specific reorder point. The EOQ provides a model for calculating the appropriate reorder point and the optimal reorder quantity to ensure the instantaneous replenishment of inventory with no shortages. It can be a valuable tool for small business owners who need to make decisions about how much inventory to keep on hand, how many items to order each time, and how often to reorder to incur the lowest possible costs.

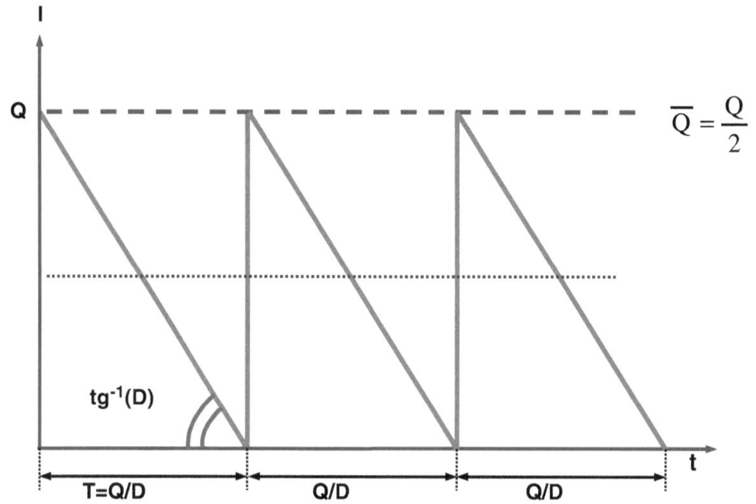

Fig. A.10 Economic order quantity logic for stock management

The EOQ model assumes that demand is constant, and that inventory is depleted at a fixed rate until it reaches zero. At that point, a specific number of items arrive to return the inventory to its beginning level.

Under the preceding hypotheses, the inventory behavior is that illustrated in the graph of Fig. A.10.

Since the model assumes instantaneous replenishment, there are no inventory shortages or associated costs. Therefore, the cost of inventory under the EOQ model involves a trade-off between order costs C_1 (any fees associated with the orders placing, such as delivery charges) and inventory holding costs C_2 (the cost of storage, as well as the cost of tying up capital in inventory rather than investing it or using it for other purposes). Ordering a large amount at one time will increase a small business's holding costs, while making more frequent orders of fewer items will reduce holding costs but increase order costs. The EOQ model finds the quantity that minimizes the sum of these costs.

The basic EOQ formula is as follows:

$$C(Q) = C_1 + C_2 = A\frac{D}{Q} + h\frac{Q}{2}$$

where C(Q) is the total inventory cost per year, h is the holding cost, Q is the order quantity, and A is the order cost (Euros per order). Breaking down the elements of the formula further, the yearly holding cost of inventory is h multiplied by the average number of units in inventory. Since the model assumes that inventory is depleted at a constant rate, the average number of units is equal to Q/2. The total order cost per year is A multiplied by the number of orders per year, which is equal to the annual demand divided by the number of orders, or D/Q.

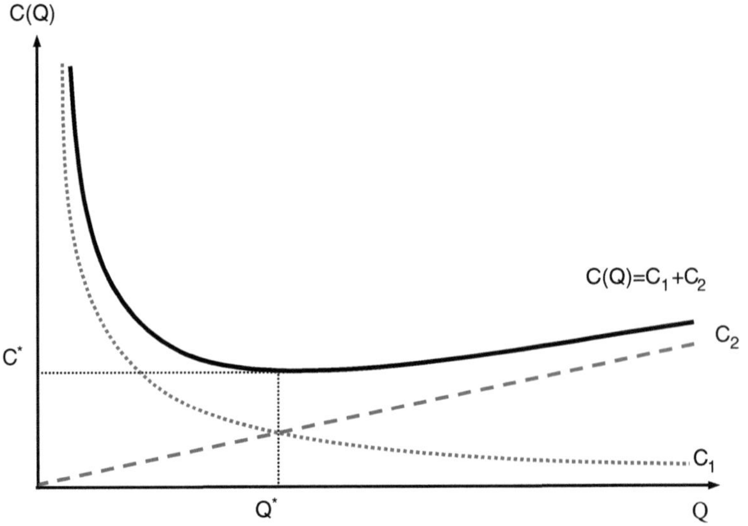

Fig. A.11 Economic order quantity curve

The Fig. A.11 graph presents the trend of the total inventory cost per year depending on the order quantity.

Considering these factors and solving for the optimal order quantity gives a formula of: $\frac{\partial C}{\partial Q} = -A\frac{D}{Q^2} + \frac{h}{2} = 0$ or $Q^* = \sqrt{\frac{2AD}{h}}$ which means $C^* = C(Q^*) = \sqrt{2ADh}$. It is easy to verify that $\frac{\partial^2 C}{\partial^2 Q}|Q^* > 0$.

Hypothesizing a determined lead time (LT), it is sufficient to define the restock point R as the products demand during the lead time: R = LT · D.

A policy of stocks control based on the restock point is extremely simple to be implemented, but the limit of the EOQ approach is due to the assumptions described. Nevertheless this model is robust, particularly if the variables to change are the costs (of which is often available only an estimate). It is logical to round off the lot dimension to practical values, still obtaining valid estimations of the C(Q) minimal value. It is possible to show that: $\frac{C(Q)}{C(Q^*)} = \frac{1}{2}\left[\frac{Q}{Q^*} + \frac{Q^*}{Q}\right]$, where Q^* is the EOQ (economic order quantity). If the lot dimension doubles compared to the economic order quantity ($Q/Q^* = 2$), the cost increase is 25 %.

A.3.5 Economic Order Quantity Variations

In literature it is possible to find many models derived by the economic lot, they are obtained "relaxing" some of the hypotheses previously seen. We quote some of them.

Decreasing Price Depending on the Order Quantities

The production or purchase unitary cost (Cu) may change in presences of discounts and economies of scale following a law like this:

$$C_u = \begin{cases} C_u^1 \ if & Q < Q_1 \\ C_u^2 \ if & Q_1 < Q < Q_2 \\ C_u^3 \ if & Q_2 < Q \end{cases}.$$

In such case the function to be minimized is:

$$C(Q) = C_1 + C_2 + C_3 = A\frac{D}{Q} + h\frac{Q}{2} + C_u \cdot D.$$

The curve of the total inventory cost is broken therefore in different branches, as it is shown in the diagram in Fig. A.12.

It is possible to apply the EOQ formula calculated in the previous section to each branch: $Q^* = \sqrt{\frac{2AD}{h}}$.

Notice how the cost minimum changes depending on the size of the order lot (Q) and it can also correspond to an extreme of a branch. To determine the optimal lot size, it is necessary to study each interval of unitary cost definition checking the extremes of the intervals. Having determined the economic order quantity for every unitary cost, the one that gives the smallest total cost C(Q) is selected.

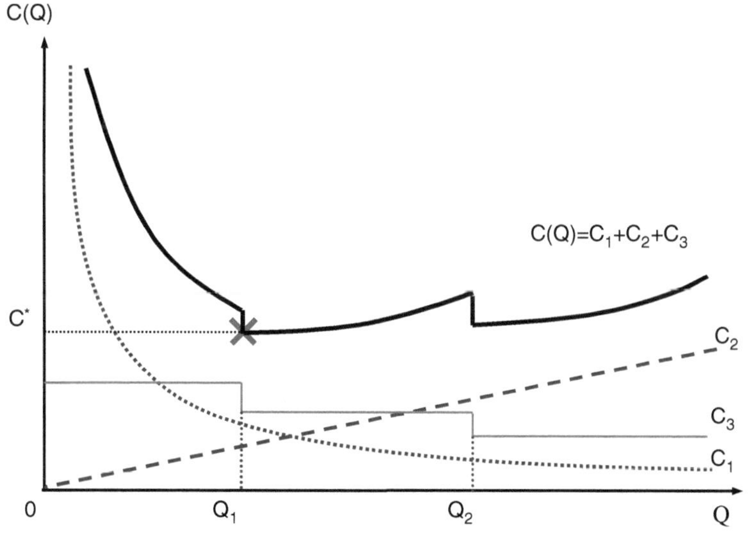

Fig. A.12 Economic order quantity curve variation depending on production unitary cost variations

Economic Order Quantity with Bounded Speed of Filling

If the producer and the store are "logistically" close to each other, the store is filled with a speed equal to the difference between the production rate (P) (number of unities produced for time period) and the demand rate (D), being obviously P > D. In such case, the time needed to produce Q unity is Q/P. Since the demand is assumed to be constant, the maximum warehouse level won't be Q but $Q\left(1-\frac{D}{P}\right)$.

The warehouse behavior is plotted in the graph in Fig. A.13.

The average warehouse lead time is equal to $\frac{Q}{2}\left(1-\frac{D}{P}\right)$; therefore the average warehouse maintenance cost between two orders is equal to: $C_2 = \frac{Q}{2}\left(1-\frac{D}{P}\right)$.

Therefore, as for the EOQ model, the function to be minimized is:

$$C(Q) = C_1 + C_2 = A\frac{D}{Q} + h\frac{Q}{2}\left(1-\frac{D}{P}\right).$$

It is possible to show that the economic order quantity in this case is $Q^* = \sqrt{\frac{2AD}{h\cdot\left(1-\frac{D}{P}\right)}}$ and $C^* = C(Q^*) = \sqrt{2ADh\left(1-\frac{D}{P}\right)}$.

These deterministic models introduced in the preceding sections are based on the assumption that the demand rate and the lead times are known quantities. Actually, both of them are unknown. In the probabilistic models (not considered in this text) such unknown quantities are modeled as random variables, characterized by a certain distribution of probability.

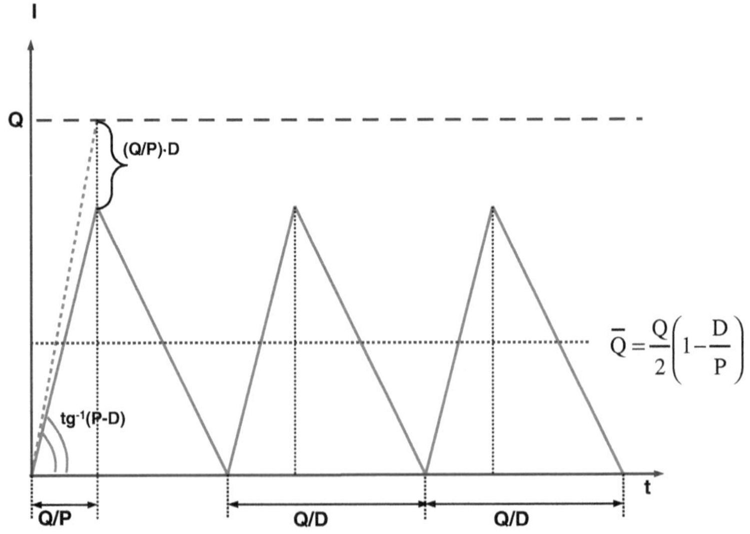

Fig. A.13 Economic order quantity with bounded speed of filling

A.3.6 Smallest Sustainable Lot Size

To determine the dimension of the production lot it is necessary to consider the effect of set-up times, in presence of production constraints. In other words, it is opportune to verify that the production is realizable in the planned lead time, not only considering some working cycle times, but also of the set-up times needed to change the machining, including the production losses due to the lot starting, whose incidence increases with the decreasing of the lot dimension.

To study the problem, in the following we describe a model based on the following hypotheses:

- N different articles or sketch-elements (for ex. A, B, C, D) must be produced on the same working machine and with the same cycle time. For each article QP_n pieces must be produced.
- The production of the various articles is organized in repetitive cycles (for example A–B–C–D A–B–C–D etc..). For each different article the lot size is proportional to the respective quantities to produce.
- The machines set-up is done before the production of each lot.
- The set-up times are the same for every typology of products and they do not depend on the size of the lots to produce.
- The machine's unavailability due to breakdowns or to programmed maintenance interventions, is ignored, assuming that it does not depend on the frequency of the set-up.

With reference to the definitions given in Sect. 3.2, we consider the following quantities:

SOT	[min./piece]	The **standard operation time** is the time needed for a machining cycle; it is supposed to be the same for all the types of articles to be produced
N		Number of different articles to be produced
AWT	[h]	**Available working time** to produce the scheduled articles
UST	[pcs]	**Unitary set-up time:** inactivity time due to the machines set-up (UST = set-up time/SOT); actually it is the number of pieces that could be produced during the set-up time
LTP	[pcs]	**Lot turnover period** in production neglecting the set-up times, defined as a number of SOT; under the hypothesis of equal SOT for all the articles, *the LTP is actually equal to the total number of articles produced during a complete lots turnover*
HVP	[pcs/h]	**Hourly virtual production** obtainable without machining changes (HVP = 60/SOT)
QP_n	[pcs]	quantity of elements of the nth article to be produced in the available working time (AWT)
QP_{TOT}	[pcs]	Total number of elements to be produced ($QP_{TOT} = \sum_{n}^{N} QP_n$)
q_n		Ratio between the programmed quantity of the n-th article and the total production: ($q_n = QP_n / QP_{TOT}$).

The percentage of set-up times on the working time is given by: $\frac{N \cdot UST}{LTP}$.

The **available productive capability** is equal to:

$$\mathrm{APC} = \mathrm{AWT} \cdot \mathrm{HVP}/(1 + \frac{N \cdot UST}{LTP})$$

If the production strategy is to produce in order to get a good products mix, it is interesting to calculate the smallest value of sustainable LTP able to guarantee the planned production program, both in terms of total quantity and mix.

Let us consider the required production capability (X), as a dependent variable. Obviously it must be X ≤ APC. We are interested in the quantity Y = smallest sustainable LTP, as a function of X.

We can define X through the following mathematical relationship: $X = \frac{HVP \cdot AWT}{\left(1 + \frac{N \cdot UST}{y}\right)}$.

Expressing Y as a function of X: $y(x) = \frac{N \cdot UST}{\left(\frac{HVP \cdot AWT}{x} - 1\right)}$.

Figure A.14 describes the y(x) behavior.

The **smallest sustainable lot size $L_{n,\,min}$** i.e. the quantity of the n-th article to be produced in each turnover period is: $L_{n,\,min} = y \cdot q_n$. The resultant value must be rounded toward the bigger unity.

First feasibility verification: in order to adopt a repetitive lot production strategy the available working time must be necessary at least to adopt a single lots production strategy (i.e. to produce first all the scheduled A products, then B, C and so on).On the other hand if it is not possible to adopt a single lots production strategy then a repetitive lots production strategy is even more impossible to realize because of the set-up time incidence.

$$\frac{SOT}{60} \cdot (QP_{TOT} + N \cdot UST) \leq AWT.$$

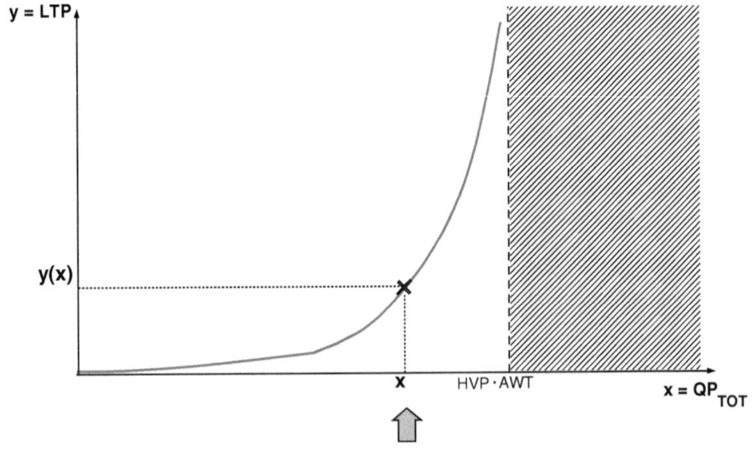

Fig. A.14 Smallest sustainable lot size trend

As it is possible to notice from the inequality above, the number of set-ups in a single lots production is equal to N.

The **number** of necessary **lots turnover (LTN)** is equal to $\frac{\frac{QP_{TOT}}{N}}{\sum_n L_{n,\min}}$ rounded towards the bigger units.

Second feasibility verification: considering the rounding off done, it is reasonable to verify the following temporal constraint:

$$LTN \cdot turnover\, duration \leq AWT \Rightarrow LTN \cdot \frac{SOT}{60} \cdot \left(\sum_n^n A_{n,\min} + N \cdot UST \right) \leq AWT$$

In order to satisfy the second feasibility condition, a practical suggestion is to reserve a part of the AWT as spare time and to introduce a slightly smaller (5–10 %) AWT in the LTP calculation. For example if AWT = 8 h it is possible to calculate LTP, introducing in the formula a TED equal to 7.5 h, reserving a spare time of half an hour to the production. As a result of the rounding off introduced, the resulting TED will be definitely slightly larger than the scheduled 7.5 h, but still smaller than the limit of 8 h.

Conclusive considerations on this theme:

1. The smallest sustainable lot size grows with the set-up time (UST) and the number of different elements to be produced. It decreases with an increase of production capacity.
2. The set-up time incidence increases with the number of different elements to be produced. For this reason it is important to adopt technical and organizational systems designed to reduce the set-up time the most possible.
3. The smallest sustainable lot approach is typical of pull production systems, while the EOQ production strategy is peculiar to push systems.

A.4 Problems Connected to the Planning Horizon

To **manage production** it is necessary to predispose the necessary resources, considering the product's sales forecast and coordinating the materials flows coming into the process of transformations and the outgoing end products.

To that purpose, it is necessary to predispose an effective strategic planning, following the criteria described in Chap. 1, in order to assign and to opportunely allocate production resources. The temporal horizon to be considered for strategic planning depends on the activity typologies: the required lead times for the product's development and for their industrialization depend on the product technological complexity and on the necessary characteristics of the manufacturing systems.

Production planning is strictly connected to strategic planning, but in different ways depending on production management logic. Its main aim is that of regulating order fulfillment, according to a client's agreement. The operational planning temporal horizon is a middle term temporal horizon (normally less than one year), even if the programs become executable month by month, week by week and day by day. The operational planning establishes purchase commitments, manpower requirements, factory operation schedules, flow of incoming materials and output of products destined to clients.

Then, the problems of planning of the production are different according to the considered temporal horizon:

- The problems of planning production capacities are handled on a long temporal horizon.
- The problems of detail planning (scheduling) are handled on a brief temporal horizon.
- The problems of control of production flows and of deliveries to clients are potentially handled in real time.

For the planning of middle-long term production capabilities it is reasonable to consider production plans for families of products (Aggregated Production Planning or **APP**). Such plans, which are used to scale the general production capability of multi-product manufacturing systems, are then disaggregated—on shorter temporal horizons—in single product plans. In such a brief-middle period temporal horizon, the most used strategies for production planning are the **MPS** and the **MRP** techniques. The **MPS** (Master Production Schedule) allows us to detail and to correct the general planning derived by the **APP**, considering the number of orders, the store's situation and the priorities of delivery to the clients. The **MRP** systems (Material Requirements Planning) allow planning the demand for direct materials related to single products, both as semi-finished components purchased from the outside or as semi-finished components produced by the same business group. MPS, MRP and Scheduling interaction is shown in Fig. A.15.

Table A.3 indicatively resumes the decisions that have to be taken in long, middle and short temporal horizons.

Different temporal horizons correspond to different **plan refresh rates**. A long term decision, which is based on long term information, should not be open to question too frequently, because a long term forecast doesn't change so quickly.

Typically long term problems are put up for discussion with annual or semestral frequency. Middle term problems are re-discussed with quarterly or monthly

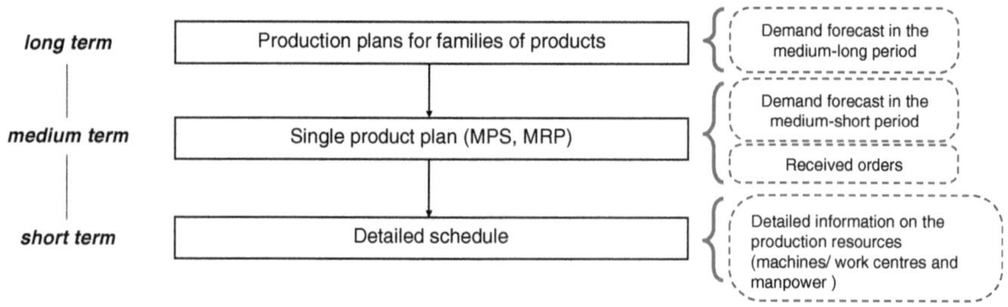

Fig. A.15 MPS, MRP and scheduling interaction

Table A.3 Problems and decision making in production planning management

Temporal horizon	Duration	More representative decisions
Long term (strategic problems)	Several years	• Marketing strategies • Make or buy policies • Products range plan • Decisions about activity levels • Choice of technological solutions • Production resources allocation • Decisions about industrial cooperation • Purchase commitment • Planning of production capabilities • Staff development plans • Quality policies • Financial plans
Medium term (tactical problems)	Annual/three-monthly	• Sales promotion • Production capability adjustment • Material supply plan • Personnel recruitment • Quality control plan • Annual budget and quarterly revisions
Short term (problems of scheduling)	Monthly/ weekly/daily	• Working times of industrial installation • Assignment of working tasks • Restocking planning • Production flows management • Production process control • Clients deliveries management • Emergency interventions

frequency. Short term problems are revised week by week and day by day, while the control of the productive flows and the dispatches to the clients is done in real time.

In the following sections the adopted techniques for working planning are described, independently of the production contexts. In Chap. 6 such matters are picked up and widened with specific reference to the automobile production.

A.5 Aggregate Production Planning

The aim of Aggregate Production Planning (APP) is to determine production, inventory, and workforce levels in order to meet a time-varying demand pattern. The APP problem pertains to strategic decision levels; therefore, APP models represent suitable aggregations of manufacturing systems. Aggregation has different facets:

- Similar products are often grouped in families; in this way the uncertainty in demand is reduced, since it is easier to forecast the demand for a family of items than for the single item (see the scheme below);

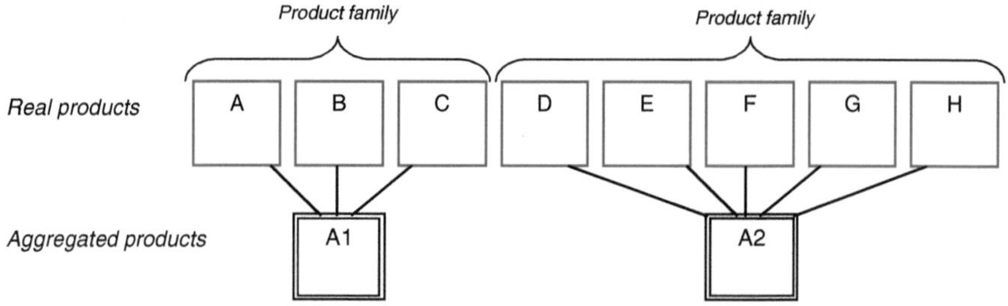

Fig. A.16 Product aggregation logic

- Time is discretized in relatively large time-buckets; this means that more than one product/family is produced during each time-bucket, but no sequencing decision is taken;
- Resources are grouped in work-centers; since we are not concerned with short-term scheduling at this level, it does not make sense to specify which machine in a work-center will carry out a certain job.

Such aggregation implies that what is obtained is not a detailed production schedule, but only a rough-cut indication of what should be done (Fig. A.16).

Aggregation has to consider the production factors which are considered more binding for the levels of activity. For instance:

- **Strategic raw materials;**
- **General and specific manufacturing systems;**
- **Workforce.**

To develop the aggregate production plan, it is necessary to have a middle-long term demand forecast, organized for families of products and for lines of production.

A.5.1 APP in the Case of Single Aggregated Product

Hypothesizing to have only a single aggregated product whose demand forecast is known in the middle-long period, for instance within six months, the cumulated demand curve is traced in Fig. A.17.

There are different strategies for planning the production in order to satisfy the forecasted demand. To didactic purpose, two opposite alternatives are now illustrated (see also diagrams in Fig. A.18).

(a) **Level strategy—using inventory as the lever:** With this strategy, a stable machine capacity And workforces are maintained with a constant output rate. Shortages and surpluses result in inventory levels fluctuating over time. In this case production is not synchronized with demand. Either inventories are built up in anticipation of future demand or backlogs are carried over from high to

t [months]	1	2	3	4	5	6
d_t [pcs/month]	30	30	120	90	60	30
d_t cumulate	30	60	180	270	330	360

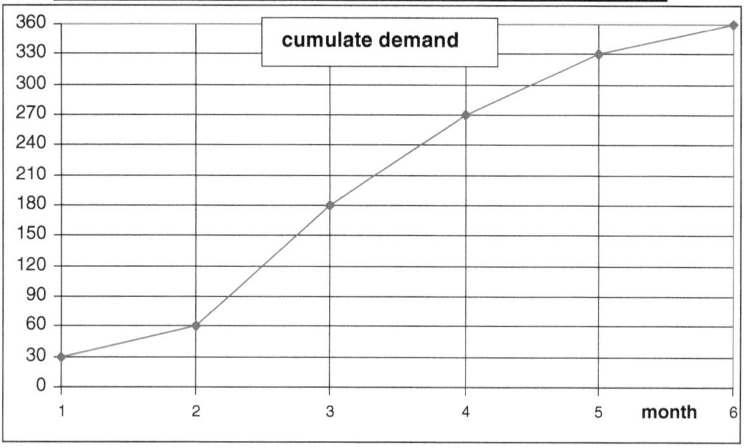

Fig. A.17 Cumulated demand curve for a single aggregated product

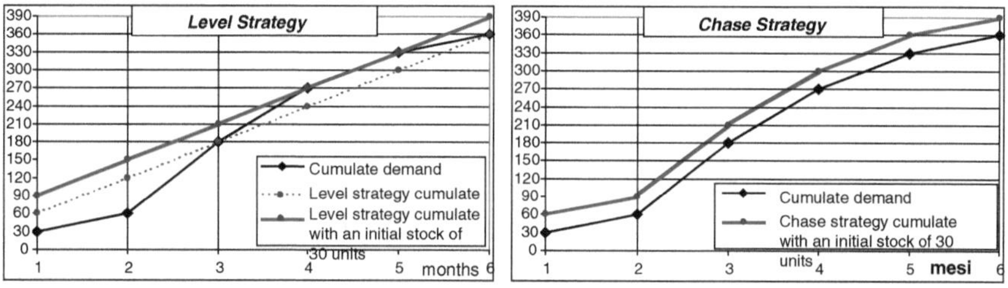

Fig. A.18 Production planning alternatives

low demand periods. Employees benefit from stable working conditions. A drawback associated with this strategy is that large inventories may accumulate and customers' orders may be delayed. This strategy keeps capacity and costs of changing capacity relatively low. It should be used when inventory carrying and backlog costs are relatively low.

(b) **Chase strategy—using capacity as the lever**: With this strategy, the production rate is synchronized with demand rate by varying machine capacity or hiring and laying off employees as the demand rate varies. In practice, achieving this synchronization can be very problematic because of the difficulty of varying capacity and workforce on short notice. This strategy can be expensive to implement if the cost of varying machine or labor capacity over time is high. It can also have a significant negative impact on the morale of the workforce. The chase strategy results in low levels of inventory in the supply chain and high levels of change in capacity and workforce. It should be used when the cost of carrying inventory is very high and costs to change levels of machine and labor capacity are low.

The two proposed strategies are two extreme admissible solutions. In reality strategies can be built combining the preceding ones. For instance it is possible to keep producing at a constant rate, but changing it in correspondence to demand peaks.

The aggregate production planning involves a preventive analysis of feasibility. For instance, after having planned the production, it is necessary to verify the resource's availability in terms of machines and workforce; if the availability is not enough it is necessary to increase it or to do overtime or sub-contracts.

A.6 Master Production Schedule

The Master Production Schedule (MPS) is a typical support for operational detailed planning, regarding the brief-middle term temporal horizon. The MPS is built from aggregate production planning (APP). The products aggregated in "families" or "lines of production" are separated into individual products. The production is separately planned, based on orders already received or on brief-middle term sale forecasts. The middle-long term sale forecasts for a single product are inevitably uncertain, because of the subjectivity of the orders. That is the reason why in the middle-long term temporal horizon, the planning is developed in an aggregate way, for families or product lines (APP).

The planning temporal terms (time bucket) are reduced as soon as the period of execution of the programs gets closer: time buckets pass from monthly bucket to weekly or daily bucket.

Summarizing, the MPS defines in a detailed way the coarse APP planning, considering every single product, in a shorter and therefore less mutable temporal horizon. For instance, if the APP foresees producing 100 units of the aggregated product P1 in a certain month, through the MPS it is decided what is the single product combination to be realized (for example 25 type A products, 45 type B and

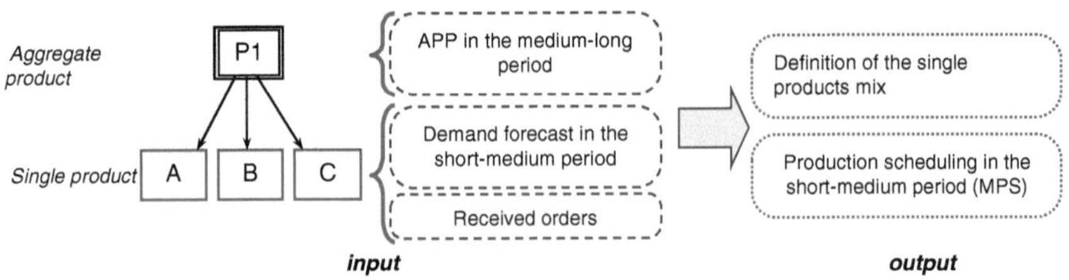

Fig. A.19 MPS logic, input and output

30 type C products), scheduling the production in weeks (see the Fig. A.19 schematization).

The MPS details and corrects the coarse planning derived by the APP, considering received orders, inventory position, delivery priorities toward the markets and requirements of single clients.

A.6.1 Structure of the MPS

The MPS is made of temporal tables reporting on every planned product. To give an idea of how it works, let us look at a simplified example:

- A demand forecast is available for 12 periods (weeks);
- Fixed quantity lots of 30 unities are produced whenever the inventory level goes under the 5 unities (fixed order quantity);
- The initial level of the store (I_0) is equal to 20 units.

With such hypotheses the MPS corresponds to Table A.4.

The second line of the table shows the brief term demand forecast for every week. The line "MPS" identifies the periods in which to effect the production and the quantities to be produced. The third line of the table points out the inventory level forecast for each period. *It is a common practice to evaluate this quantity at the end of the period, i.e. after all the requirements have been satisfied.* According to what stated above the inventory level forecast is calculated as:

Inventory level forecast (t) = inventory level forecast (t−1) + MPS(t) − demand forecast (t)

This program is feasible, if the demand doesn't overcome the productive capability related to every considered period (t).

Actually the MPS is more complex than it seems, since it must take into account the received orders and the deliveries that can differ from the forecasts. For instance if during the first time periods some unexpected orders are received, the production planning has to be "corrected", modifying the MPS. If the MPS is not updated, it is possible to produce less than what would be necessary or more than the demand, uselessly burdening the warehouses.

Table A.4 MPS structure example

t [weeks]	1	2	3	4	5	6	7	8	9	10	11	12
Demand forecast	5	5	5	5	5	5	15	15	15	15	15	15
Inventory level forecast (I_0 = 20)	15	10	5	30	25	20	5	20	5	20	5	20
MPS	–	–	–	30	–	–	–	30	–	30	–	30

A.7 Material Requirements Planning

The final product demand is considered **independent**, because it comes from factors, such as the market demand, which are external to the production process. On the contrary, the subcomponents demand is considers dependent, because it is tightly connected to the finished product demand. For the elaboration of the Material Requirements Planning (MRP) each product is broken down in terms of subcomponent and relative base materials, according to what established by the **bill of material** (BOM), as we have seen in the precedent Sect. 2.3.

Both the two policies of stocks management (periodic review, fixed order quantity) can result unsatisfactory, when the market demand is affected by short-middle term variations since they assume a constant market demand. On the other end the stocks management policies discussed have the advantage of being simple and doesn't necessarily require the use of information systems.

To avoid this assumption in the sixties the Material Requirements Planning (MRP) system has been introduced.

The MRP systems allow to plan the material demand, both for semi-finished products that are purchased from the outside, and for semi-finished products produced inside the company. The MRP systems are very useful for the stock management in order to define the quantity or the time periods in which to purchase or to produce. The MRP has a short brief temporal horizon (generally few weeks).

The logic of the MRP is a production planning logic:

- Which supposes an unlimited production capability;
- Which is directed towards a stocks reduction.

The production capability constraint is not directly taken into account, but it is considered through its principal effect, which is the creation of delays, which force to arrange the production or purchase orders with enough advances in comparison to the demand. The advance is equal to a **lead time** a-priori fixed. For instance if during the *n-th* time bucket there is a requirement for 20 unities of a certain product and the fixed lead time is equal to 3 periods, then the order is planned for the time bucket *n-3*. Such process is called **lead time offsetting**.

The MRP data structure is constituted by a chart, called MRP record. The temporal horizon is divided in periods called time buckets that at the very beginning were monthly. Thanks to the evolution of the processors time buckets passed from being monthly to weekly and then daily. In the MRP tables (presented in the following) the following information are reported period by period:

- **Gross requirements (GS).** They express the demand forecast during each time bucket. The gross requirements of the end-products correspond to the MPS planning, which gives a projection of the future production. The gross components requirements depend on the received demand of the aggregated product.

- **Scheduled receipts (SR).** They correspond to the orders (of production or purchase) already placed, that are expected to be satisfied at the begging of a certain time bucket.
- **On-hand inventory (OH).** It is the inventory level at the end of every time bucket, after that all the orders have been satisfied. The inventory level for the MRP is not said to be equal to the available inventory level: in fact, all the dead stock of defective or expired elements (in case of materials affected by deterioration) and of elements assigned to other production plans must be excluded.
- **Net requirements (NR)** are obtained from the gross requirements, applying the aforesaid "net-ification".
- **Planned order receipts (OP).** Purchase/production orders, whose accomplishment is programmed in correspondence of every time period; they are obtained from the net requirements considering the lot sizing criterions adopted.
- **Planned order releases (OR).** Purchase/production orders placed in each considered time bucket; they are obtained from the planned orders considering the lead time offsetting.

The MRP logic proceeds recursively beginning from the end-products, going down along the bill of material and "exploding" the requirements of each end-product, also considering the necessary times to put into production the orders in all the levels of the integrated production process.

The necessary information for the construction of the MRP is:

- *The BOM* of all the end and intermediary products;
- *The MPS* related to the single products, that is the respective gross requirements;
- The initial inventory and safety stocks situation;
- *Lot sizing* rules used (quoted in the following);
- Situation of the *scheduled receipts*, related to the orders of end-products and sub-components which have been issued in the preceding time buckets;
- Information about the *lead times*.

Let consider the example in the underlying tables. An end-product P1 needs the production of a subgroup P2; the production of a subgroup P2 in turn needs the production of two components P3 (according to the Bill of Material). 10 unities of P1 and 20 unities of P2 are available in the warehouse; besides a delivery of 20 unities P2 is expected at the beginning of the third week. The lead time for P1 is one week, two weeks for P2 and three weeks for P3. The lots sizing is banal for P1 and P2; the lot size correspond to the net requirements (this rule of lot sizing is known as *lot-for-lot*). For P3, it is hypothesized, instead, that the orders have a constraint: the size of the order must be a multiple quantity of 50 unities (*fixed order quantity*).

This information is usually gathered in a resuming table that is defined as *item master file*, as shown in Table A.5.

Starting from these data (see Table A.5), the net requirements are calculated and the placing of the orders is planned. The arrows in Table A.6 point out the advantage necessary to satisfy the demand.

Table A.5 Item master file

Item master file					
Product/ component	Initial inventory level (OH_0)	Safety stocks	Lead time	Scheduled receipts	Rule of lot sizing
P1	10 unity	–	1 time bucket	–	Lot-for-lot
P2	20 unity	–	2 time buckets	20 unity during the time bucket 3	Lot-for-lot
P3	–	–	3 time buckets	–	Fixed order quantity (Q = n • 50)

Table A.6 Net requirement calculation example

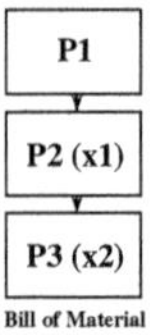

End-product P1 (Lead time= 1 week)

T [weeks]		1	2	3	4	5	6	7	8
(GR) gross requirements = MPS						50		60	
(SR) scheduled receipts									
(OH) on-hand	OH_0=10	10	10	10	10	0	0	0	0
(NR) net requirements						40		60	
(OP) planned orders receipts						40		60	
(OR) planned orders releases					40		60		

sub-component P2

t [weeks]		1	2	3	4	5	6	7	8
(GR) gross requirements					40		60		
(SR) scheduled receipts				20*					
(OH) on-hand	OH_0=20	20	20	40	0	0	0	0	0
(NR) net requirements							60		
(OP) planned orders receipts							60		
(OR) planned orders releases					60				

* notice how the delivery of 20 unities of the under-component P2, attended for the period 3 could be delayed of one period, not to uselessly burden the warehouse.

sub-component P3

t [weeks]		1	2	3	4	5	6	7	8
(GR) gross requirements					120				
(SR) scheduled receipts									
(OH) on-hand	OH_0=0	0	0	0	30	30	30	30	30
(NR) net requirements					120				
(OP) planned orders receipts					150				
(OR) planned orders releases		150							

As it can be observed, the MRP allows elaborating a planning starting from the established moment of the end-product delivery, in order to define the moments in which to effect the placing of the orders and the relative production period of each component elements, specifying the relative quantities.

If the lead times are correctly estimated, the MRP logic allows to optimize the stocks and the WIP (work in process), since it synchronizes the production of the necessary components for the assemblages. In this way the materials become available in the moment in which they are necessary. Once placed, the planned orders become operative and they are turned into scheduled receipts. To launch an order with too much advance is unadvisable since the stocks can burden the warehouses and it is difficult to modify an operative order to conform itself to the changes of the MPS.

To counterbalance the demand uncertainties, the MRP has to be updated with a certain frequency, introducing the occurred variations. In the requirements calculation it is possible to plan the necessary safety stocks; in this way the net requirements are not produced not when the inventory is empty, but when it goes under a certain threshold. The safety stocks allow to cope with the peaks of demand, and to make up for to possible delays.

When more components of the same type are necessary to assemble a parent product, as for P3 in the preceding example, the number of parent product orders must be multiplied for the quantity of necessary sub-components to assemble a single parent product. Such number can be overestimated to consider of the real performance of the production process in order to compensate the number of defective parts. For example, to forehead of a net clean requirement of 100 pieces, an order of 110 pieces can be placed, assuming that around 10 pieces will be defective. Naturally this approach is discussable, since it assumes that the defect is noticed after the produced lot has been stored into the warehouse.

In the shown examples it is assumed that the products requirements and the dead stocks in inventory are quantified in numerical terms and not in terms of weight or volume.

A.7.1 Lot Sizing in the MRP Systems

The rules of lot-sizing in the MRP packets have been object of researches. Nevertheless there are not sure "recipes" for the adoption of a rule of lot-sizing. There are varying quantity rules for which the inventory level is null at the end of the horizon of planning (unless a safety stocks level has been specified); this is the case of the rule lot-for-lot, adopted for the products P1 and P2 of the preceding example. For product P3, the fixed order quantity rule is adopted instead, with the consequence that at the end of the horizon of 8 weeks, there is still a dead stock of 30 pieces. It follows a brief description of the most common rules of lot sizing.

- The **fixed order quantity** (FOQ) rule a-priori fixes the order quantity. Sometimes such policy is made necessary by the technological process that furnishes the lots in fixed quantity, or from the possibility to order only a fixed quantity (or multiple of a given quantity) from the suppliers. The fixed quantity to be ordered can also be calculated applying the rule of the economic order quantity (EOQ). This requires to determine the average demand per period basing on the requirements. Hypothesizing a fixed cost (A) equal to 100 and an unitary cost for a dead stock (h) equal to 1 and an average demand equal to 120/8 units per period, the application of the EOQ formula gives:

$$Q^* = \sqrt{\frac{2AD}{h}} = \sqrt{\frac{2 \cdot 100 \cdot (120/8)}{1}} \approx 55.$$

- The fixed order period (FOP) rule assumes that the period between two consecutive orders, is a-priori fixed. The fixed order period defines the number of periods (time buckets) whose requirements are covered by the order.
- The simpler lot sizing rule is the **lot-for-lot** rule, in which net requirements and orders coincide. In this way the costs of inventory are minimized, at the fixed costs' expenses. The rule is applied as illustrated in the preceding example, for the products P1 and P2. There are some possible variations of this rule, for example with the addition of a smallest sustainable lot size.

A.7.2 Application of the MRP Logic to a More Complex Case

Let's analyze another example of the MRP application. The example refers to two bill of material for two end-products P1 and P6, as in Fig. A.20 scheme.

To assemble a piece P1, two units of P2 and one of P4 are needed. Let notice that P3 is used for producing both P2 and P6. It would be wrong to proceed with the calculation of the requirements for the various products in an arbitrary order. When the calculation of the gross requirements of a product is performed, it is necessary to make sure that his "parent" products orders has already been planned. To guarantee this, a number of levels are attributed to every product starting from the end-product. P3 is involved in different levels (lev. 2 for P1 and lev.1 for P6), but it is assigned to the lowest level; this is known as **low level coding**. *The requirements of the product P3 are aggregated because it would be unreasonable and expensive to effect more different "micro-orders" (and therefore to build more*

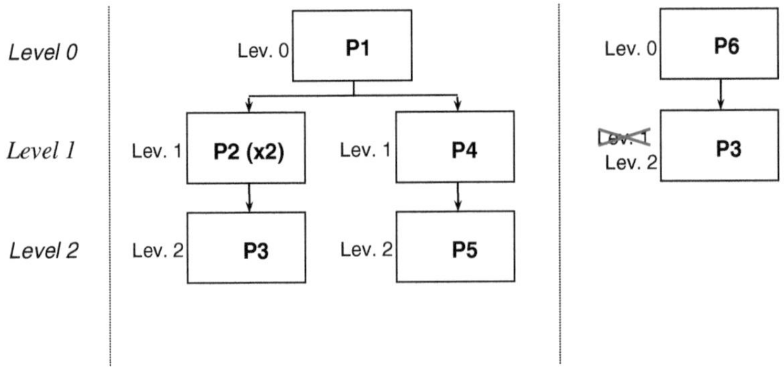

Fig. A.20 MRP application on a complex case example

Table A.7 Item master file example 1

Item master file						
Product/ compon.	BOM level	Initial inv. lev. (OH_0)	Safety stocks	Lead time	Scheduled receipts	Rule of lot sizing
P1	0	30 units	–	1 period	–	Lot-for-lot
P2	1	–	–	2 periods	–	Lot-for-lot
P3	2	40 units	20 units	1 period	–	Fixed order quantity (Q = n(50))
P4	1	–	–	1 period	20 units during the period 3	Lot-for-lot
P5	2	–	–	2 periods	–	Fixed order period (3 periods)
P6	0	–	–	1 period	–	Fixed order period (3 periods)

t [weeks]	1	2	3	4	5	6	7	8
MPS (P1)		20				40		20
MPS (P6)				30	10		10	

MRP) for a same sub-component. The data to build the MRP are shown in the underlying **item master file** (Table A.7).

The calculation of the requirements proceeds, level for level, as illustrated in the MRP Table A.8, that is substantially self-explanatory.

In the moment in which the demand for a certain product cannot be satisfied thanks to the scheduled receipts (SR) or thanks to the on-hand (OH), the net requirement (NR) of every product is equal to the gross requirement (GR). More precisely: $NR(t) = GR(t) - (OH(t-1) + SR(t))$.

Once calculated the net requirement and considering the selected rule of lot-sizing, it is possible to plan the orders.

It is worth to underline as the gross requirements for P3 are given by the sum of the planned orders releases for P2 and P6. Besides, the P3 MRP shows the effect of the safety stocks; in the period 3, despite a requirement of 100 pieces and an availability of 40, the net requirement is of 80 pieces and not 60. To facilitate the calculations, the safety stocks can be systematically subtracted from the on-hand; in this case the initial availability of P3 is only of 20 pieces (40–20) rather than 40.

Inside the MRP it is possible to correct the scheduled receipts, bringing them forward or backward, to make the MRP admissible or to optimize the inventory. There are two types of rectifications:

(1) **Expediting**: a request to anticipate the scheduled receipt in comparison to the planned period in order to be ready in case of a demand peak;
(2) **Deferring**: a request to delay the scheduled receipt in comparison to the planned period, not to burden the inventory in the periods of low demand.

The *expediting* and the *deferring* are rectifications that cannot always be granted, both in the case the supplier is internal or external, because it may involve

Table A.8 MRP table example 1

End-product P1

t [weeks]		**1**	**2**	**3**	**4**	**5**	**6**	**7**	**8**
(GR) gross requirements			20				40		20
(SR) scheduled receipts									
(OH) on-hand	OH_0=30	30	10	10	10	10	0	0	0
(FN) net requirements							30		20
(OP) planned orders receipts							30		20
(OR) planned orders releases						30		20	

End-product P6

t [weeks]		**1**	**2**	**3**	**4**	**5**	**6**	**7**	**8**
(GR) gross requirements					30	10		10	
(SR) scheduled receipts									
(OH) on-hand	OH_0=0	0	0	0	10	0	0	0	0
(FN) net requirements					30	10		10	
(OP) planned orders receipts					40			10	
(OR) planned orders releases				40			10		

Sub-component (Lev. I) P2

t [weeks]		**1**	**2**	**3**	**4**	**5**	**6**	**7**	**8**
(GR) gross requirements						60		40	
(SR) scheduled receipts									
(OH) on-hand	OH_0=0	0	0	0	0	0	0	0	0
(FN) net requirements						60		40	
(OP) planned orders receipts						60		40	
(OR) planned orders releases				60		40			

Sub-component (Lev. I) P4

t [weeks]		**1**	**2**	**3**	**4**	**5**	**6**	**7**	**8**
(GR) gross requirements						30		20	
(SR) scheduled receipts				20 →		**20***			
(OH) on-hand	OH_0=0	0	0	~~20~~	~~20~~	0	0	0	0
(FN) net requirements						10		20	
(OP) planned orders receipts						10		20	
(OR) planned orders releases					10		20		

*** scheduled receipts delayed of 2 periods (deferring)**

Sub-component (Lev. II) P3

t [weeks]		**1**	**2**	**3**	**4**	**5**	**6**	**7**	**8**
(GR)gross requirements (OR_{P6}+ OR_{P2})				100		40	10		
(SR) scheduled receipts									
(OH) on hand	OH_0=20+20	20+20	20+20	20+20	20+20	30+20	20+20	20+20	20+20
(NR) net requirements				80		20			
(OP) planned orders receipts				2·50		1·50			
(OR) planned orders releases			100		50				

Sub-component (II lev.) P5

t [weeks]		**1**	**2**	**3**	**4**	**5**	**6**	**7**	**8**
(GR) gross requirements					10		20		
(SR) scheduled receipts									
(OH) on-hand	OH_0=0	0	0	0	20	20	0	0	0
(NR) net requirements					10		20		
(OP) planned orders receipts					30				
(OR) planned orders releases			30						

Table A.9 MRP results example 1

Product/ component	Quantity [unity]	Period of release	State	Annotations
P1	30	5	OK	(OR) planned orders releases
P1	20	7	OK	(OR) planned orders releases
P6	40	3	OK	(OR) planned orders releases
P6	10	6	OK	(OR) planned orders releases
P2	60	3	OK	(OR) planned orders releases
P2	40	5	OK	(OR) planned orders releases
P4	20	5	delay	(SR) schedule receipts delayed of 2 periods
P4	10	4	OK	(OR) planned orders releases
P4	20	6	OK	(OR) planned orders releases
P3	100	2	OK	(OR) planned orders releases
P3	50	4	OK	(OR) planned orders releases
P5	30	2	OK	(OR) planned orders releases

some changes in the planned production. If the rectifications are feasible then the MRP is admissible. The rectifications are generally resumed in a table similar to the following. If such rectifications are not realizable, the MRP can result inadmissible and therefore it has to be planned again.

Table A.9 contains the results of the MRP.

A.7.3 The MRP Nervousness

The rules of lot-sizing that use different quantity product orders (such as the lot-for-lot rule or the fixed order period rule) are affected by a phenomenon called nervousness. Small variations in the input MPS can have remarkable effects on the requirements computation. Here it follows a typical example of nervousness, in which the plan becomes inadmissible because of a slight variation of the gross requirements (Table A.10).

This plan is admissible, but if the gross requirement of P1 suffers a light change, passing from 24 to 23 units in the period 2, the MRP of P2 it changes notably (see Table A.11).

The reduction of the product P1 gross requirement causes "nervousness" because the production plan changes notably and produces an urgent order. The order of 47 units, attended for the second period, is delayed of four periods.

There are different approaches to avoid the nervousness. The adoption of rules of fixed quantity lot-sizing allows to face the variations of small entity, even if with an increase of the inventory costs. Quantity varying lot-sizing rules are generally preferred for the end-products, not to burden too much the warehouses; at the same time for the sub-components fixed quantity rules are adopted, in order to contain the nervousness, whereas some troubles would be produced. It is possible to adopt

Table A.10 Item master file and MRP table example 2

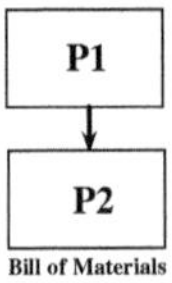

ITEM MASTER FILE				
product / component	**On-hand (OH_0)**	***lead time***	**Scheduled receipts**	**rule of *lot sizing***
P1	28 units	2 periods	-	*fixed order period* (5 periods)
P2	2 units	4 periods	14 units during the period 1	*fixed order period* (5 periods)

End-product P1

t [weeks]		**1**	**2**	**3**	**4**	**5**	**6**	**7**	**8**
(GR) gross requirements		2	24	3	5	1	3	4	50
(SR) scheduled receipts									
(OH) on-hand	OH_0=28	26	2	13	8	7	4	0	0
(NR) net requirements				1	5	1	3	4	50
(OP) planned orders receipts				14					50
(OR) planned orders releases		14					50		

Sub-component P2

t [weeks]		**1**	**2**	**3**	**4**	**5**	**6**	**7**	**8**
(GR) gross requirements		14					50		
(SR) scheduled receipts		14							
(OH) on-hand	OH_0=2	2	2	2	2	2	0	0	0
(NR) net requirements							48		
(OP) planned orders receipts							48		
(OR) planned orders releases			48						

Table A.11 MRP results example 2

End-product P1

t [weeks]		**1**	**2**	**3**	**4**	**5**	**6**	**7**	**8**
(GR) gross requirements		2	~~24~~ **23**	3	5	1	3	4	50
(SR) scheduled receipts									
(OH) on-hand	OH_0=28	26	3	0	58	57	54	50	0
(NR) net requirements					5	1	3	4	50
(OP) planned orders receipts					63				
(OR) planned orders releases			63						

Sub-component P2

t [weeks]		**1**	**2**	**3**	**4**	**5**	**6**	**7**	**8**
(GR) gross requirements			63						
(SR) scheduled receipts			**14***						
(OH) on-hand	OH0=2	2	0	0	0	0	0	0	0
(NR) net requirements			47						
(OP) planned orders receipts			47						
(OR) planned orders releases		**47****							

*** scheduled receipts delayed of 1 period (deferring)**
**** order delayed of 4 periods**

strategies of diversified management of the planning temporal horizon. In the short term temporal horizon the starting MPS is generally not changed; in the middle term temporal horizon the MPS can be changed after specific analysis and authorization; the MPS is changed at one's pleasure in the long term temporal horizon instead.

A.7.4 Critical Analysis of the MRP Systems

The MRP system aims to reduce the stocks through the calculation of the net requirements and placing the necessary orders the later it is possible to, subtracting the lead time to the delivery dates. It is really the lead time the core of the matter; if it was possible to have an exact estimate of the lead time, the MRP system would allow to produce what it is necessary when it is necessary. Nevertheless the lead time can be considerably greater than the time required for the product manufacturing; since it depends also:

- From the queue times;
- From the times required for the materials movement;
- From the possible need for reworking due to the presence of defects;
- From the lot subdivision, due to the presence of important set-up times.

These dead times are impossible to be exactly foreseen. Therefore it is natural "to inflate" the *lead time* in comparison to its average value, in order to be sure not to remain without materials. Great l*ead times* cause high-levels of WIP. During the dead times, the materials and the semi-finished products are steady, stuck in the warehouses or in some intermediary buffer. The presence of long *lead times* creates a particularly perverse vicious circle. In presence of long *lead times*, it is necessary to adopt a long term horizon of planning. But a long term horizon of planning causes a greater level of uncertainty, relating to the input data (MPS), which in turn causes an increase both of the *lead time* and of the safety stocks. A partial solution to this problem is to reduce the time buckets passing from weeks to days, or even to job turns. This doesn't resolve the capacity problem, but at least it reduces the possible amplification of the *lead times*.

About the Author

Marco Gobetto was born in Rimini (RN, Italy) in 1967 and took his master's degree in Electronic Engineering in 1993 at Politecnico of Turin.

He currently holds positions as Manufacturing Training Unit Manager at FIAT Sepin S.c.p.A. and Professor at the Faculty of Automotive Engineering. He has taught courses in both Production Management and Production Processes, Safety, Organization and Management.

After having complete his University studies, he began his professional career at FIAT Auto S.p.A. in the Manufacturing Engineering Department, as part of the Style Design Manufacturing Analysis Team. From there he proceeded to the position of Body in White Process Engineer, contributing to the production launches of two new cars in Italy and then coordinating industrialization activities in Poland for a new body in white model.

In 1998 he moved to Ferrari S.p.A. as Body in White Plant Manager at the "Carrozzeria Scaglietti" facility in Modena.

In 2000 he left FIAT to assume responsibility for the Manufacturing Engineering Department at Carrozzeria Bertone S.p.A. in Grugliasco (TO, Italy). In two years, three new models of vehicles were launched and, for the first time in Bertone history, second and third shifts in production were activated.

At the beginning of 2002 he rejoined the FIAT Group at ISVOR FIAT S.c.p.A. where he specialized in Total Productive Maintenance and Continuous Improvement training programs, following on IVECO and FIAT Powertrain sectors. He subsequently coordinated international training initiatives for a joint venture in India between FIAT and TATA. In 2007 he assumed responsibility for the Manufacturing, Quality and Product Development Training Area that, at the end of 2008, became the central Manufacturing Training Unit in the new training organization in FIAT, thereby contributing to all initiatives worldwide both in FIAT and Chrysler LLC with a major focus on World Class Manufacturing, the Corporate Production System.

M. Gobetto, *Operations Management in Automotive Industries*,
Springer Series in Advanced Manufacturing DOI: 10.1007/978-94-007-7593-0,

MIX
Papier aus verantwortungsvollen Quellen
Paper from responsible sources
FSC® C105338

If you have any concerns about our products,
you can contact us on
ProductSafety@springernature.com

In case Publisher is established outside the EU,
the EU authorized representative is:
Springer Nature Customer Service Center GmbH
Europaplatz 3, 69115 Heidelberg, Germany

Printed by Libri Plureos GmbH
in Hamburg, Germany